GET REAL

A POSITIVE SOLUTION TO CLIMATE CHANGE

DAVID MUNSON, JR.

Performance Publishing Group
McKinney, TX

ISBN:
Softcover: 978-1-956914-24-5
Hardcover: 978-1-956914-25-2

CONTENTS

Acknowledgement.......xi
Foreword.......xiii
Dedication by Each of Us – It is time to "Get REAL" on climate change!.......xvii

CHAPTER 1: THE GET REAL PROGRAM.......1
Strategic Planning Includes Natural Gas.......2
It Is Time to Invest.......4
The Real Story.......5
The Value of Collaboration.......10
The Get Real Program as a Leader for Change.......12
Essence of The Get Real Program.......12
The Get Real Program Priorities.......15
The Big Payoff.......18

CHAPTER 2: A TREASURE: EARTH'S CARBON CYCLE.......21
Now, Let's Talk Carbon.......22
The Carbon Cycle.......25
Carbon Dioxide (CO2) Levels.......27
The Soil Solution.......28
How to Make Carbon Work for Us and Not Against Us.......29
The Get Real Program Believes in USING NATURE for CHANGE.......36

CHAPTER 3: THE RICHES: ROCK DUST AND REMINERALIZATION OF SOIL.......39
Without Rock Dust..........40
Life in the Lifeline... Rocks and Rock Dust.......41
The Glacier Connection to Our Soils.......44

The Miracle of Soil Life Revitalized Through Remineralization . . 45
Rock Dust to the Rescue Through Remineralization 47
Technology to the Rescue 53
A New Environmental Approach 54
Rocks Matter! 56
Carbon Sequestration Solution: Rock Dust 57
The Big Win of Basalt and Rock Nutrition for Soils and the Future 58

CHAPTER 4: MAGNIFICENT WONDERS IN BIOCHAR 59
Why Biochar? 60
The History of Biochar 62
The Making of Biochar 63
Regenerating with Biochar 66
Biochar and Water 67
The Biochar Investment 67
Necessity of Innovation 69
Empty Arguments 77
Let's Globalize 78
Biochar Is the Answer 78

CHAPTER 5: THE MIRACLE: SOIL LIFE 81
The Soil and Climate Change Connection 83
A Brief History of Soil 83
Microorganisms and Our Soil 85
The Chemistry Within 86
The Soil and Your Health 88
Man's Role in Our Soil 91
Value of Properly Investing in Our Soil 95
YES, There Are Solutions... by The Get Real Program 96
The Time to Begin Is NOW 104
Trust in the Farmer, Not Those with Lined Pockets or the Researcher in a Lab 105
Imagine a Healthier, More Prosperous World for ALL 106

CHAPTER 6: THE MARVEL: GRASS AS A CARBON SOLUTION 109
Threats to Our Grass 111
The Complex Connection from Grass to Cow... and Its Critical Role in the Carbon Cycle 113
Holistic and Regenerative Grazing of Grasslands 123
High-Intensity Prairie Grass Management 126
The Next Phase... Biogas 132
Vulnerability in Our Food Supply 134
Innovative, Yet Doable Solutions 136

CHAPTER 7: RICHES IN REFORESTATION AND THE MAGNIFICENCE OF OUR EVERYDAY TREES AND DESERT SHRUBS 141
Love for the Forests 143
Solutions and More 148
The "Who Pays For It?" ... in Climate Change 156
Creative Solutions from Learned Minds 158
Let's Be a Champion of Our Trees and Forests 160

CHAPTER 8: WONDERS IN OUR WATER: OCEANS, REEFS, WETLANDS, AND MORE 161
Remineralization Is Required 164
Critical and Interesting Solutions for Our Environment and Our Water 166
Imagine It... Carbon Sequestration Advances with Biorock 173
Opposing Views 176
The "How" Behind the What Within Ocean Solutions 177
A New Nautical Tomorrow 178

CHAPTER 9: LET'S GET REAL WITH SMART ENERGY OPTIONS 181
A Place in Time 182
Options for Change 184
Efficient Land Transportation 204

More Efficient and Cleaner Overseas Transport 206
The Danger of Relying Too Much on Long-distance Electricity 206
The Politics and Energy Policy 207
The Get Real Program Knows Better 214

CHAPTER 10: SOLAR ENERGY 215
Solar Based Energy Production 219
Aluminum 229
Efficiencies and Sustainability 230
The Politics and Energy Policy 231
Solutions Exist 232
The Get Real Program 233

CHAPTER 11: HOW TO DO IT... PERSONAL ACCOUNTABILITY 235
The Options.... The way that we see it is that we only have two pathways for change 237
Let's Talk REAL Impact 240
A Global Program 250
Start Today By... 257

CHAPTER 12: FREEDOM IN GENERATING CHANGE FOR REAL IMPACT 259
Who Can Solve for the BIG Issues? 264
The Plan for America Serves as a Model for the World 268
You may wonder how we can have the greatest impact? 270
Investment in Our Soil and Other Regenerative and Remineralization Efforts, Including Restorative and Regenerative Agriculture 271
Let's Focus on Our Soil 271
Let's Maximize Our Grass 274
Let's Clean Up Our Forests 278
Let's Stop Ignoring Our Waters... Oceans, Reefs, Wetlands, and More 279

Carbon Sequestration as a Solution 280
Climate and Energy 281
A Global Program 283
A Prosperous Future 283
YOUR Call to Action Starts HERE and NOW!!! 287
Reference List 291

ACKNOWLEDGEMENT

I wish to thank all the many people who helped make this book a reality. I stand on the shoulders of giants in many fields who laid the foundation helping to create my climate program that offers hope for the future.

Great thanks to Caroline Slaton for creating the graphics and designing the cover. The staff of Performance Publishing helped greatly with rewriting and editing. Scurry Johnson of HSJ Consulting Inc. and the Get Real Alliance staff did great editing and made improvements as well. Joanna Campe and her staff at Remineralize the Earth were a great help with the chapter on rock dust.

FOREWORD

This book is worth reading regardless of what you think about allegations of impending catastrophic climate change.

Despite false allegations by the liberal mainstream media that there is an alleged "scientific consensus" that catastrophic climate change is happening, responsible scientists reject climate hysteria. Ordinary Americans are divided on whether climate change is really an existential threat. Most rightly suspect climate hysteria is being promoted by government and media for political reasons—to advance "green socialism" at the expense of economic prosperity, jobs, and freedom, while growing government power.

However, all Americans, of all political persuasions, care deeply about the environment. Everyone wants to reduce environmental pollution and protect Mother Nature. Everyone wants governments, industry, and individuals to be responsible stewards of natural resources and the Earth. Everyone wants a clean and healthy biosphere for ourselves and future generations.

This book argues for lowering carbon dioxide (CO2) levels by solving problems other than climate change that we all can agree are far more certain and harmful to life on Earth. Instead of focusing on reducing emissions as the sole project which can slow the rise of carbon dioxide, it focuses on increasing natural carbon sequestration so that atmospheric levels drop while we still use oil and natural gas efficiently. We can all agree we need to preserve topsoil, grow more mineral rich food, stop forest-die-off, shrink deserts and increase ocean health.

And if the climate hysterics are right about impending climate catastrophe, the strategies outlined in this book will solve that problem too—without "green socialism." Having had conversations with David I can testify that he is very rational and makes a strong case for his program. As an expert on the threats to our electric grid with https://emptaskforce.us/

I agree with his warning that promoting an all-electric future is very risky without a focus on protecting the grid.

I hope you will read the book and take action by supporting the organizations that are discussed in the book. We can "save the planet" without sacrificing prosperity and liberty.

– Dr. Peter Vincent Pry

Dr. Pry is Executive Director; Task Force on National and Homeland Security, a Congressional Advisory Board dedicated to achieving protection of the United States from electromagnetic pulse (EMP), cyber-attack, mass destruction terrorism and other threats to civilian critical infrastructures on an accelerated basis. emptaskforce.us. Former Chief of Staff: Congressional EMP Commission (2001-2017).

"David Munson Jr. (Founder of Get Real Alliance) is one of the most generous people I have ever known. His care for the world and people is inspirational. He spends countless hours researching how to best help others and has dedicated his life to make the world a better place."

– Carrie Headington
Founder, Good News Initiative

I first met David Munson, Jr. in front of my butcher shop late one evening when he showed me some of his grass-fed beef. He claimed that he had somethinq that all the other producers who had come to my door did not: marbled, tender beef, and he did.

David produced beef for our butcher shop for more than a decade and became a friend. He has always had a vision to produce high-quality local beef on a large scale. When he initially started talking to me about carbon sequestration, I knew that he had landed on a recipe that could make local beef production profitable and sustainable. I have learned over the years that when David presents a far-fetched idea or theory, if I wait three years, I can read about it in publications; in 5 years, it is widely implemented, and in 10 years, it is a practice. I hope this process takes hold and gains acceptance within that timeline. If that happens, we might just save ranching and the world.

Matt Hamilton
Founder/ CEO
Local Yocal
McKinney Texas

DEDICATION BY EACH OF US – IT IS TIME TO "GET REAL" ON CLIMATE CHANGE!

R: Responsibility; E: Educate; A: Action; L: Lasting.

It is our collective Responsibility to Educate ourselves and take Action to create Lasting impact on our environment.

Few areas of study are as controversial and layered in such depth as climate change. This book is here as your trusted resource to break down the volumes of material on the topic and tell you what matters most, provide reasonable and innovative solutions, add transparency to corrupt and misguided information, and guide you on how ***you too can create impact***.

There is a tremendous urgency for us to educate ourselves and "Get REAL" on what is happening in our environment. No longer do we have the luxury of burying our heads in the proverbial sand. Disinformation and distraction from the real issues have wasted time and money that could be directed towards real solutions that have true and sustainable impact.

There are too many voices out there with propaganda that serve their ego and wallets over progress and change. As commercial ex-rancher and farmer, engineers and students of agriculture, we at THE GET REAL ALLIANCE have a front seat perspective of what matters most on the topic of climate change and our environment. It is our intention to give you the power to make a difference for your future, as well as the generations yet to come. The best time to get involved was yesterday, so let's get to it!

The purpose of this book is to help us to "Get REAL" on what is happening in the soil we walk on and in the air we breathe in order to shed light on the most popular of topics… climate change and, most importantly, the health of our planet. You will be enlightened on these complex topics through the lens of engineers by training, an inventor, a rancher, and a student of agriculture. The content offers a deep-dive analysis regarding the real-world impacts of rising carbon dioxide (CO_2) levels and other more critical issues involving our environment and sustainability. It gets real about smart energy options (i.e., renewable energy) and how America has the power to create a pathway to be revitalized for a more prosperous future.

The Get Real Alliance (GRA), which I founded recently, has developed a program ("The Get Real Program" or "The Program") with real solutions that are doable and, more importantly, affordable. You may find it unbelievable that these positive solutions are not already part of the proposals that address rising CO_2 levels and energy planning. The Get Real Program shares many years of accumulated research and study of the challenges we are facing surrounding the sustainability and health of our planet.

There are a lot of financial reasons that make people dependent on this topic being spun one way or another, which has created a lack of trust and a level of confusion among the public. This is why we were compelled to get the true on-the-ground facts out there into the public's hands from the earnest viewpoint of an eagerness to maximize life and sustainability here on Earth. It is my intention that this book serves as a starting point for many to learn and invest in real solutions that will benefit all who are living and yet to be born.

In this book, getting REAL means accepting that breakthroughs in technology are needed, but it is the *application* of proven techniques and practices around the world that will move the needle on climate change. This book offers some innovative ideas and inventions, but primarily it attempts to combine and break down a treasure trove of proven knowledge that you can use as a foundation for action. It compels and details positive action to solve multiple serious issues that will allow America to become the leader of sustainable environment-saving solutions and thereby to thrive.

As a matter of emphasis, however, I feel that it is necessary to be forthright and earnest to counter some very unrealistic and harmful proposed remedies for climate change. This book is intended to, in an unbiased way, provide information through a unique lens on topics that are not always included in discussions on our environment. The political division and negativity among climate activists could severely damage our future and modern way of life without solving the rising CO_2 levels. It is unfortunate that our environment has become a politicized lightning rod, but the money is flowing and that can bring out the bad actors. We should all be working together for the collective good of saving the planet.

There is a better way than scare tactics to garner support and get attention on the right issues and solutions for our environment. Sadly, many attempt to shock people with dire climate forecasts that only paralyze action. If the picture painted is so dire and beyond repair, then people may just decide to give up. The promotion of extremes and negative viewpoints can have devastating consequences for people, including loss of hope, short-sighted living, mental illness, and a lack of purpose. People need to be inspired, not scared, for change to be most impactful.

Without proposing a positive solution-oriented pathway towards a more prosperous future, the efforts will be stifled as that of a few, rather than the many. Great leaps can be made in America, leading the way for those of fewer resources and lack of a positive perspective. You may wonder what the alternative could be... confusion and misdirection will cause wasted efforts, or worse, be counter to the goal of positive environmental impact.

Perhaps, people will become somewhat dismissive by the outrageous claims, as they also hear countering messages of how man has managed to thrive, to some extent, despite the challenges. These messages may be easier to absorb and accept, as the alternative problem-focused messages can feel debilitating. Consider Thomas Malthus, an influential 18th century economist, who incorrectly painted a dark future of starvation and deprivation - that was overcome by innovation. There are, however, serious issues and threats that need to be addressed and people need to unite for the common good with positive solutions.

This book is full of positive, realistic solutions to very urgent and real issues surrounding us today. The Earth is not static but is both upgrading and degrading over time. In this book, you will learn how nature itself holds keys to sustainability of the environment. For instance, healthy soil, a miracle of life, is a critical part of the solutions moving our nation forward to a better way of living and breathing.

There are some very real and simple solutions within nature. Unfortunately, most climate activists are dismissive of solutions that do not line their pocketbooks. This book will break it all down and demonstrate how many scientists, working in a lab and not in nature, get stuck on false theories and fail to see value in the more reasonable solutions that are provided within this book.

Possible solutions are intermingled throughout this book. The most important solution is sequestering carbon in ways that are proven, fundable, and solve other problems as well. Another includes several innovations and ways of doing things that may prove to be well-adopted and beneficial to soil fertility, CO_2 sequestration, lower energy costs, and better more sustainable farming practices. We can do our part, but we also need the federal government to take action by restoring public forests to health and stopping forest fires. This program, unlike others, doesn't require us to give up or phase out the use of oil and natural gas to make the world carbon-negative in a short time. This is unlike the consensus on climate change that involves only trying to slow the rise at extreme sacrifice. The Get Real Program creates sustainable impact with little sacrifice. The Program has a bold vision to make America carbon-negative, not just slow our carbon emissions for little impact.

Another big difference is that our program is intended to be funded by charitable donations for the most part, instead of government taxes and mandates. If we don't make the hard decisions and take effective and timely actions, the alternative is either continued desecration or bigger government. We can make the world carbon-negative with the support of those who think climate is a priority, without punishing the poor and those who don't think it is a worthwhile issue – as most Americans seem to feel, based on their purchasing decisions. We can only do this with bold action and investment. Almost everyone wants to make sure we have adequate food that is sustainable, so many may choose to support The Program for that reason.

I hope you will find this book uplifting and promote its concepts to friends and politicians because there is personal urgency within these solutions. My hope is that your time and effort spent studying and absorbing the material in this book will have multiplied your inspiration to become part of my movement for positive change towards a bountiful future for all. Hey, maybe by providing the best evidence in this book, I will enlighten even the worst of the naysayers and darkest of climate activists to join my mission supporting my vision for positive change.

This book was conceived and mostly written before COVID-19 became a pandemic that is altering our world to an unknown extent. The very natural and affordable solutions to the climate crisis contained in this book cannot be lost in the distraction of our crisis of today. If we delay action, we will only move from the COVID-19 pandemic to another more real and devastating crisis that has only increased in intensity during the waiting period. With the world in uncertain times, it is critical to be diligent in planning for the future, because it will be here before we know it. The clock is ticking!

As a former profitable grass-fed beef rancher who continues to be a student of agriculture, I look forward to sharing my life's work with you. I remain involved in the meat industry through investments, as local food production is still an area of great importance to me. I am very concerned about posterity and, as an active investor and inventor, I seek to make the world a better place. Many of my viewpoints were established because of personal experiences, but other sources have provided validation and a foundation to my work here. If this book inspires you to learn even more or you feel the need for personal corroboration, I encourage you to read the books and authors which are referenced in the book.

I also recommend you check out the Get REAL Alliance website (https://getrealalliance.org) for up-to-date information and sign up to receive the newsletter, which will provide lots of valuable information and resources to assist you in getting and staying REAL… supporting your continued education and efforts. Contact us at thegetrealalliance@gmail.com.

I implore you to keep an open mind throughout the material in the book, as you may have differing viewpoints on some of the topics but may align on others. Even where you may differ on solutions, you will gain value and may find common ground with the foundation of the issues. Each invention and analysis stands on its own merits but are all intended to work to make things better. It is my dream that the labor of love that it took in bringing my life's work to you will inspire you and spark new, innovative ideas for progress and action within you.

Let's lead by example and work together to make this world a better place. Let's dedicate ourselves to "Get REAL" today for a better future for generations to come! - David Munson Jr. - Founder, The Get Real Alliance.

It is our collective Responsibility to Educate ourselves and take Action to create Lasting impact on our environment.

CHAPTER 1

THE GET REAL PROGRAM

"Everything is theoretically impossible, until it is done."

– Robert A Heinlein

The Get Real Program (aka "The Program") solves the direst of environmental challenges that America and the world have ever faced, and the time to act is now. Few sources share detailed truths about what we are facing in regard to the greatest threats to our everlasting ability to sustain healthy living. So many activists are out and about stirring up scary stories on rising air temperatures and CO2 levels. The immediate focus should honestly be on drastically slowing soil erosion, the growth of deserts, deforestation, and ocean decline.

If we solve these real problems, the CO2 level in the air will naturally go down while we still use oil and natural gas in the most efficient ways. Let's work together and make lasting positive environmental change not only possible, but our reality. How, you ask? Get behind The Get Real Program and we can do just that!

Do you want to ensure a healthier and more sustainable environment today and in the future for generations to come? We all do! The question is *how* we do it. To accomplish this goal, we must shift the discussions in the media around popular topics like solar photovoltaic and things we can see and feel in our everyday lives, like rising air temperature, to the real problems, innovations, and solutions that lie in less apparent topics that we have had unique views of within nature.

Unfortunately, the CO2 debate and propaganda on solar photovoltaic as "the" solution are just a distraction from the real underlying issues that need immediate attention. Importantly, The Get Real Program addresses how we now live in a very demineralized world that makes it nearly impossible to lower CO2 levels unless we create change through efforts like remineralization to further carbon sequestration. **Carbon sequestration, which is fixing and storing carbon in forms other than CO2 gas in the atmosphere (so effectively removing CO2 from the atmosphere), is key to real change and having a sustainable impact on the real issues affecting our environment today.**

We are all being led astray from focusing on the real threats to a prosperous environmental future. The Get Real Program has simple, natural solutions to address our desire for sustainability and progress through natural gas and more. Natural gas is far and away the cleanest, most abundant fuel and we need to use a lot more of it. In order to solve for our current circumstances, we must take an inventory and utilize all of our natural resources available to us according to a strategic plan.

STRATEGIC PLANNING INCLUDES NATURAL GAS

Natural gas is not the problem; instead, it is part of the solution. It has the lowest percentage of carbon of any natural fuel. It can be easily created renewably and biologically, but it is abundant geologically as it is one of the most common compounds on Earth behind water, sand, and rocks. Natural gas optimization coupled with global carbon sequestration funded by private, responsible actions of charities will do more for the environment than misguided efforts that keep carbon emission high by preserving coal for political reasons while benefiting the coal unions.

You have heard of the phrase "no pain, then no gain." The environmentalists took this phrase literally when they devised the current solutions. And this is where we see the divide. Everyone wants the ability to breathe better and can appreciate the feeling we could get from our positive individual impact on the environment for generations to come. We all want the opportunity to feel good about what we are putting into our bodies and those of our children. To get there, we need a comprehensive pathway that is more realistic than the proposals of the day and that is discussed among the "experts." This is where The Get Real Program brings affordable and actionable solutions that are proven for today's problems and those that are certain to come due to inaction.

Workable solutions that don't create chaos and a detrimental impact on how we go about our daily lives are key. We don't want to create new problems or take a back seat to innovation while looking backwards to decisions of the past. The Get Real Program's vision is for the future, but it is built upon what history has shown to us.

We are all intelligent and can see how data can be manipulated to provide the conclusion any funder desires. This is where real-life experiences and the solutions outlined in The Get Real Program come in and shed light to provide REAL solutions without significant pain in how we go about our lives. No other solutions out there can manage the complex and comprehensive issues, as they are too often limited, one-sided, or have conflicting agendas to line the pockets of the messengers.

The agenda of The Get Real Program is to restore health in our precious Earth, regardless of what your feelings are about what may be impacting our environment and to what level. The debate should be over for all, as the time to act is now. Let's work together to lead the way, instead of being led by special interests. Let's follow the clues the Earth is providing, not the flashy dollars of special interests.

When you are on the right side of things for the right reasons and you are qualified, confidence will follow. As The Get Real Program is shared by you and others, our efforts will grow. They will grow because transparency will show their value. Wow, it is a beautiful sight to see transparency in our climate. Do you see it?

IT IS TIME TO INVEST

Just like with our bodies, no matter our level of health, it is important for the Earth that we maintain constant balance. This means taking in the right level of nutrients and taking full advantage of our natural resources without abusing our system in any way. No longer can we ignore the Earth's signals that we need to invest back and tend to our soil, deserts, forests, and oceans.

Any worldly thing deteriorates over time, if not tended to, and our natural resources are no different. The Get Real Program has been devised from a vision for real impact that will be less painful to our everyday lives, yet allow us to move on to other greater challenges of the day once we begin implementation. One way or another, investment will be required. Either we will start today, or we will pay a bigger price down the road when trying to clean up from the passiveness of the past. In fact, with millions of tons of topsoil eroding every day, there is no time to waste in making needed changes to agricultural practices that are at the heart of rising CO2 levels.

Does this sound good to you? Do you want to help? Then let's get down to the real matters and create change together. We need you—yes, YOU—as the proposals in The Get Real Program will require a lot of education, investment, and footwork to elevate the right solutions for real progress. Let's get started.

THE REAL STORY

The truth is... the atmosphere in totality is very complex and it is functionally impossible to model accurately as a whole. There are too many intricate connections and variables to map out all of the realms of possibilities accurately into one fixed conclusion. At best, components can be measured. The GRA thinks we can all agree that there are alarming trends when looking at the components of change we are seeing around us.

As a farmer for many years and native-Texan, the founder has seen the impact of a crisis as recently as the winter of 2021 when Texas had record low temperatures for a week, which led to tremendous power outages, food shortages, deaths, and more. Shortly after this anomaly in record low temperatures, the very warm temperatures wiped out any effect that week had on the average. In fact, with Alaska warmer than Dallas, the warm temperatures in the huge area within Alaska more than offset the cold week in Texas.

This event demonstrated that it is not the averages that matter, but extremes and timeliness. Yet, the extremes are not typically the focus. This is why we have seen climate activists change their language from "global warming" to "climate change," as the atmosphere wasn't cooperating with their warming language.

The truth is... the Earth has a wide array of temperatures all the time and we need to be concerned with something more important, which is the amount of photosynthesis done on Earth. This is impacting us in incredible ways, yet no one is mentioning it. Photosynthesis is shrinking due to the growing deserts, urban encroachment, and dying forests. This is an area where we need to see real change. Photosynthesis can increase with shrinking deserts, with better forest and ocean management, and with warmer temperatures generally due to less plant freeze, and longer growing seasons.

The truth is... rising air temperature, which has been a major focus of concern from environmentalists, has very little impact on the rate at which ice and snow melt. In fact, raising the temperature of a region one or two degrees on average through the year has no impact, because it's below freezing most of the year where ice is located. Even raising the temperature from 40°F to 41°F has almost no impact on melting. This is true because direct solar radiation (sunlight) exceeds heat transfer from the warmer air (in shade) by several orders of magnitude. Anyone who has gone skiing or has been around snow knows that it melts even when the air is below freezing due to the immense power of the sun. Blocking the sun lowers the air temperature quickly as heat is radiating out to space all the time at a much lower rate than the sun radiates heat in.

The truth is... the climate is very complex and the role of water vapor in the climate is just so immense compared to CO_2. This fact doesn't get much credit: dry lands, like deserts, don't retain moisture or have plants emitting water vapor. Therefore, increasing deserts lower the amount of water vapor in the world. In fact, deserts are net radiators of heat out to space because they don't have water vapor to hold heat in. Also, due to the role of migration from desertification, the UN is estimating the displacement of some 135 million people by 2045. Shrinking deserts and increasing water retention in the soil along with increased photosynthesis will change the world for the better but may also actually increase average global temperature while CO_2 levels go down! We need moderate temperatures, which is what water vapor gives to the climate.

As you know, if you've been to the desert, at night it gets quite cool. The Romans actually made ice in the desert by placing a cup of water in a hole and covering it with a polished shield in the daytime and opening it to the sky at night.

The truth is... even though there is urgency in phasing out harmful coal as quickly as possible, many climate activists want to preserve coal. You can see this by their efforts to block cleaner, much lower-carbon-emission natural gas power plants. Despite this opposition, America's carbon emissions have gone down due to half of the coal-fired powerplants being shut down. We have the ready supply of natural gas and ability to quickly phase out the remaining plants to further lower America's carbon emissions and make it possible for more use of renewables, like solar photovoltaic. Lowering carbon emissions while saving money is a win-win and makes it easier to make America carbon-negative by boosting sequestration.

A coal power plant takes 12 hours to start making power, so it has to stay in operation all the time to provide reliable power when intermittent renewables, such as solar photovoltaic, are out of commission. Natural gas power plants can start and stop quickly, which is ideal to back up renewables like solar photovoltaic. We need a kilowatt of natural gas power for every kilowatt of solar photovoltaic to have reliable power that we can count on, but the special interests fighting natural gas are hiding this fact from the public.

For now, it is important to know the real threat to the environment is coal burning, and we need to try to stop that as fast as possible. In our rush to manage this very real threat, we cannot create more issues for America. It is important to create practical change in a way that mitigates collateral damage. The solution is to increase natural gas power generation, while taking steps to secure jobs and create stability for those in states currently dependent on coal or coal production for their budgets.

The truth is... the current proposals to use wind, solar photovoltaic, and batteries to provide power are totally unrealistic. The solar photovoltaic industry is telling an incredible lie about the true availability of solar photovoltaic power, day in and day out, and all around the year. We should all know or accept that solar photovoltaic power output is half as much in the winter as it is in the summer. The peak power time is far shorter because the days are so much shorter. Days of cloudy weather mean no solar photovoltaic power.

The fact is that there is not an affordable or realistic way to have batteries to last for days of cloudy weather along with all the extra solar photovoltaic panels needed to charge them.

This alone is enough to realize why solar photovoltaic cannot be our only solution, although it should be part of the solution. To posit that you're going to have significant power at all times when you need it from solar photovoltaic technology is just an incredible deception. It is just not possible to live in the innovative and technologically savvy world that we do today with solar photovoltaic power alone. And why would we when we have other solutions for environmental health and reparative efforts for this Earth?

The occurrence of the Texas winter storms of 2021 is as far back as we have to go to see how wind is also not the solution. During the storms, many of the wind turbines in Northwest Texas (see U.S. EIA data for Feb 14-16) froze up and failed to supply power, and most importantly heat, during a time of record lows. This event was devastating, and even deadly, for those without backup generators.

The truth is... carbon is at the heart of life on Earth. Life can exist without oxygen, but not without carbon. The CO2 levels have fluctuated greatly over time, as there are many ways for the carbon cycle to be altered. The carbon cycle should not be at the root of our fear, but utilized at the root of our solution for real environmental concerns and to drive prosperity.

Two major dynamic stores of carbon are the ocean and the soil. When looking at soil's role in the carbon cycle alone, it is important to know that it is larger than man's burning of fossil fuels. Deforestation and demineralization of the soil are where the real issues lie as to some of the most harmful effects on our carbon cycle. It is fascinating to see the complexity of the carbon cycle and The Get Real Program's use of nature for restoration on the current concerning trends.

The truth is... the problem of soil degradation and loss is actually a core reason why CO_2 levels remain high, along with the state of the ocean and marine wetlands. The marine wetlands alone have the potential to soak up much of man's carbon emissions if they were remineralized and revitalized instead of being degraded. The soil degradation alone will destroy man's ability to survive long before 2100, if nothing is done to stop the current trends.

Soil degradation is not as interesting to talk about, but we are losing topsoil at an astronomical rate. **The U.N. Secretary-General António Guterres recently stated that over 20 billion tons of fertile soil are lost every year.** The time to act is now. We can begin implementation of The Get Real Program and begin reversing the impacts sooner than imagined with other more costly and burdensome theoretical plans. GRA looks forward to sharing with you in later chapters about all of the tremendous benefits you and generations to come will receive by the replenishment of our soil.

The truth is... many of the changes proposed by environmentalists will have unintended consequences. As we shared in the introduction, the Founder was a farmer with unique experience with the land, air, and water. Many scientists can only speak in theory and provide solutions according to historical data or limited experiments that are not real life. Unfortunately, as the environment changes and shifts, the theories based upon historical information and data may already be stale or no longer applicable.

The complexity of our environment has been his life's study, so as the environmental factors shift, models and expertise continue to elevate. This innovative and visionary approach takes into account all of the intertwining of factors that make our environment function from a place of true experience, not theory. The Get Real Program advances with the changes and complexities in the environment, unlike the siloed solutions of solar photovoltaic or other interests that are not comprehensive and fluid to the environment's leadership.

THE VALUE OF COLLABORATION

We need to work beyond agendas and transparently with one another for a real solution. The pathway will require a key group of experts in multiple disciplines to conquer our biggest challenges. The world does not have the luxury of being deaf or blind in any given area in order to have real impact. We all need air and food to survive, so we are in this together, whether we like where we are or not. Therefore, we must work with the so-called perpetrators of climate to create holistic change. We can see the solutions, but we must also know that we can solve far more than we can currently imagine by working together. Yet, while unity is difficult under current political discord, unity of planning and of action are keys to real change!

The Get Real Program takes a step forward in a bold direction for real change. The conclusions are backed up by a broad base of experienced intellectuals who are generally outside of the conventional mindset and research structure. GRA will share many resources throughout this book to provide more depth in key areas of interest. The Program and other solutions within this book are not tied to the status quo or special interest groups.

The only special interest that drives The Get Real Program is an interest in doing what is right for our natural resources. We need to stop what we are doing to the land and sea, as well as change our harmful practices in an affordable and practical way. The Program is comprehensive and presents the outline for how to save the world, but it still needs you to make it a reality. We do ***not*** need to accept contentions that our current ways of life are so grossly unsustainable and harmful that we must drastically change many of our ways of life now.

There are solutions without extreme measures that push back our innovative progress. For example, the brilliant holistic grazing expert Allan Savory calls for very different management of grasses than currently exists. He suggests more grazing as wild herds, not less, as was done before man changed things. This is a great example of one part of a solution to the issue of rising CO2 levels and it isn't tied to destroying the affluent way of life of the developed countries, as proposed by the proponents of the Paris Climate Accord.

The types of restrictive pathways now being tossed around leave no restriction on rising Chinese and Indian coal emissions that may far outpace the harsh restrictions of the Accord. The Get Real Program alternatively supports being more efficient and working towards long-lived, reliable, sustainable sources of power from a global perspective. The Program steps away from the media advanced solutions of large-scale solar photovoltaic arrays and horizontal axis wind turbines that harm our wildlife by killing birds.

Individuals can create change that is critical to success of The Program too, by getting local about power and essentials. This means creating more local options for food and power, rather than the modern way of adding huge distances to the attainment of food and power. The transportation creates much waste and leaves regions dependent and vulnerable to disruptions.

There are many wise experts out there, but you often have to get away from the mainstream supported institutions to find them as the current paid-research model takes away independent thinking. You see this with the rejection of remineralization by many with chemical industry ties or activists who cling to sharp restrictions on energy use as a near-religious belief. We have a wild situation where leaders proclaim a need for harsh restrictions for the masses while they use private jets and live in large houses.

We cannot ignore different perspectives and views in order to be the most impactful. We know listening is a rare attribute of the many these days, but that is exactly how the environmental crisis will be resolved. Do you want to learn more? OK, then let's get to The Get Real Program.

THE GET REAL PROGRAM AS A LEADER FOR CHANGE

Natural environmental solutions exist for our depleting resources! The Get Real Program can have the greatest impact when approached holistically, as the environment is a delicately structured system of organization requiring a multitude of advances at the same time. The Program is a comprehensive one that requires us to be realistic, instead of having false notions that you can have reliable power from an intermittent source or that not investing in our resources and environment will have a positive outcome. It is an affordable, realistic plan to make America carbon-negative at a very affordable cost and to do it quickly, not over decades. The Program will also restore health and wealth to America, but we all have to get on board supporting The Program's vision for real impact.

ESSENCE OF THE GET REAL PROGRAM

Increased carbon sequestration is the key to lowering CO2 levels and there are several ways to accomplish this:

1. Remineralization of land and sea should be a major part of The Program as without minerals, life can't sequester carbon and our world is very demineralized.
2. Large-scale production of biochar from dead trees and crop waste prevents carbon from going back to the air and creates biologically active sinks for nutrients that now leach away to the ocean.
3. Increasing holistic and regenerative agricultural practices that increase carbon sequestration on farms and grasslands.

If concerned citizens and companies step up and fund the needed efforts, we can make the world carbon-negative while we still use oil and natural gas. The long-time charity www.remineralize.org can easily grow to fund major remineralization with support. Please make your contribution to that growth by signing up for monthly support.

You can read a lot more in the following chapters on solutions and insights that you may not have thought of before. The Program also includes being sensible about emissions by pursuing efficiency and smart energy options that don't get the attention they deserve as they aren't coming with a low price tag but offer reliable, long-lasting power time instead of the intermittent, unreliable power of solar photovoltaic! When we think about sustainability, we must remember that the best way may be more expensive than an intermittent and unreliable one.

All can embrace solutions in The Get Real Program that clean up our environment, no matter where we stand on the topic of climate change. We should step away from the label of "climate change" and focus on real change... the kind of change that we can all buy into... the kind of change that has real impact and sustainability. If we are not the leader, Russia or China will certainly step in and take advantage of the costs of the status quo in America and the rest of the world.

Competing countries would like nothing more than to see a financially crippled America with energy prices rising to very high levels. Proposals to ban hydraulic fracking in America are great for Russia. The consequences can be dire, as American production would quickly fall and the world would be in an energy shortage, driving prices up to new highs. Domestic gasoline prices could double or triple within a year.

When looking at solutions, there is nothing better than carbon sequestration, but the effort takes investment. The Program understands that global sequestration funding voluntarily supported by the masses for the good of all would be much cheaper when compared to banning fracking! We can all be part of the solution by investing in our future. Do we want to wait and have to defend America's sustainability and leadership, or do we want to take action now? We vote for NOW! Do you?

Just like our bodies, the Earth wants to self-correct and has a natural desire for balance. Our efforts will serve the natural journey back to Earth's balance by using natural solutions and incentives to create real change. The Earth will self-correct and heal as we invest in remineralization and more. We just have to guide the way and provide the support to get it back on the right path. We will do this through taking advantage of God's gifts by channeling natural oil and gas to the best use, while preserving our advancing way of life.

The Get Real Program has a unique perspective based upon intimate knowledge of the carbon cycle. The Program will exploit the patterns within the cycle for greater atmospheric vigor and natural regeneration of our Earth's components. The carbon cycle can be enhanced by increased photosynthesis, both on land and in marine wetlands, by increasing carbon sequestration. This will ensure America is a carbon-negative country and will boost rural income dramatically, while increasing food security and quality.

Fortunately, America has been seeing huge decreases in carbon emissions, which have been tied to natural gas solutions. Unfortunately, America only impacts a small percentage of carbon emissions throughout the world. However, our potential to sequester CO2 is perhaps double our carbon emissions. This means that we can have a positively disproportionate impact on carbon emissions around the world by harnessing this resource.

Unlike China, America is rich in fertile land and forests but still has deserts that sequester almost no CO2 at all. In fact, there's hardly any photosynthesis in the desert. For example, an acre of fertile American farmland or American woodland sequesters an enormous amount of carbon compared to what occurs in many countries, where much of their land is desert or sterile. A global effort is critical, but we can start today in America and grow as we prove our positive progression. America can lead.

China has a real problem with a huge desert as part of its country. They can make some improvements in their agriculture and the way they manage things, but The Get Real Alliance is going to have to focus primarily on operations in the other countries that have a greater potential to sequester carbon to create real change. Of course, China needs to quit using so much coal and try to phase that out as fast as possible. This will require them to stop building new coal plants, but they are not focused on the same incentives as The Get Real Program currently, so no change is expected without incentives for change. There are things we all can do, but we need a global program to really be effective. This will require big players with conflicting interests buying into The Get Real Program. To promote interest in The Program, real monetary incentives have to be available to counter and offset the motive to profit by competitive advantage when America moves to greater resource management and other innovative efforts related to the GRA Program.

THE GET REAL PROGRAM PRIORITIES

A few of The Program's priorities are laid out below with the solutions being described more thoroughly in the later chapters of this book.

Critical to The Program's priorities would be to invest in our soil and other regenerative and remineralization efforts. It is vital that our soil be managed in a way that brings our soil back to a state of being nutrient-rich instead of trending towards sterile desert. By solving for our tremendous soil being devastated by man's practices and more, we can also see huge advantages in our own health and that of the environment. Soil can be a tool for carbon sequestration as it is restored to health using natural processes.

Another one of The Program's priorities includes rapidly building new natural gas generation to replace coal as fast as possible. This priority can be funded by investors for an affordable way to have cleaner air and would benefit the environment for all. The Program also takes into account the impact of the lost jobs from the coal mining industry. Now is not a time when America can afford to lose jobs and let's face it, the coal miners were just doing their job.

If we have a reasonable carbon tax including a significant tax on coal, it could fund many areas within The Program. There are two options for change: one that is government-focused and the other is private. It is clear that we all must act... and now! Instead of leaving the coal miners' families and communities destitute and unemployed, it would only take a small amount of the funding provided to charities to also invest in the miners by paying out their pensions and letting them retire with dignity in support of their families' sustenance. It is not lost on us how the challenges to coal mining towns in the Appalachian region have led to increased incidences within America's opioid crisis of today. Too many lives have been lost from the devastation of the coal mining industry - overnight in certain areas. We cannot afford—neither is it humane—to allow this to be exacerbated by further financially unsupported closures.

Additionally, The Program would provide resources for the miners to transition their skills into other areas by providing job training opportunities, job search support, and other benefits that Corporate America provides to displaced workers transitioning out of a company. The payoff of more highly skilled workers out in the workforce would outweigh any costs. The government has some responsibility as to solving issues their regulations have created, but we also like the idea of private organizations creating positive change in this area to take us further.

There are many ways to reach the goals of The Get Real Program, but we prefer the path that creates the least strife and pain to America. We don't know about you, but it is our hope that Americans privately step up for big changes, rather than leaving it entirely to the government. If private companies and charities create big changes to meet the needs of our planet, we can avoid relentless and ineffective regulation and less burdensome taxation.

The Get Real Program's priority would also be to encourage more realistic and better forms of renewable energy that have fewer negative side effects. Of course, it might cost a little more initially, but in the end, it would provide America with options that offer high-quality power and reliability, such as geothermal, biogas, algae, vertical axis wind turbines and solar thermal processes. The alternative of horizontal axis wind turbines is deadly to birds and flying creatures. Meanwhile, solar photovoltaic produces very unreliable power. We also need new generation nuclear plants and other new technology.

Funding has been channeled to the solar photovoltaic industry, which has created a treasure trove for China. China provides the majority of supplies for the solar photovoltaic industry, so even if this were an effective solution for big change, it again leaves America dependent. The Get Real Program's goal is to advance America's sovereignty in energy, while also advancing our leadership in atmospheric vigor!

The Get Real Program is looking for investment that can be re-invested in restoration and regenerative agriculture, not an empty funding stream of large tax credits to support a "false witness" like the solar photovoltaic industry. A private global fund to advance carbon sequestration can do more for our environment than the government programs on solar photovoltaic. Showing that data is ignored by special interests and demonstrating that solar photovoltaic is not THE core solution is important to maintain the economic success and growth potential we have.

Due to the mercantilist practices of the Chinese, solar may appear to come with a cheap price if you base that on the maximum nameplate power rating, not the actual much lower average output that may be near zero on many days. It may support the goals of globalists efforts, but it is not a be-all and end-all for climate change and for our needs within the deteriorating components of the environment. It will not solve for the larger impact areas that are creeping up on us, like soil erosion, growth of deserts, deforestation, and ocean decline. You need only to look to your local farmer for validation on these fronts.

It is the proposition of The Get Real Program to utilize solar thermal power plants that use natural gas heat as a backup when the sun doesn't shine, which will provide quality electricity with minimal fossil fuel use. They might cost a little more than a solar photovoltaic array, but by being able to produce power 24/7 and having other benefits, they are actually cost effective. It is a falsehood to pursue something saying it is the cheapest solution using its advertised rated capacity when it hardly ever makes that much power.

As another priority, The Get Real Program would implement efforts to encourage efficiency. It does not serve anyone's interest or needs to be so wasteful. Efficiency is one of the biggest gains and a low-hanging fruit for real impact. You can do this by easy modifications that are not detrimental to how we go about our daily lives. Just change a lightbulb over to LED each time one goes out or proactively, move to better installation practices by builders (for instance, more insulation and reflective roof coatings), as well as moving to other improved building and home designs. It likely does not even have to be said that increasing car efficiency could also create huge gains for our environment.

There are many solutions within the material in this book that can create real impact.

THE BIG PAYOFF

Imagine a world where you know the air you are breathing is getting cleaner every day.

Imagine a world where diseases are slowing in diagnosis, due to the cleaner environment.

Imagine a world where the food you taste is much richer and sweeter.

Imagine a world where America could sustain our nutrition needs in a very secure manner, knowing the food we eat is not laced

with pesticides or other poisons that less reputable companies and countries may allow to enter our food supply.

Imagine a world where the poor can afford healthier food rich in nutrients, not just those who can pay for the luxury of organic food free of unnatural pesticides.

Imagine a world where we are all engaged in preserving our natural resources for sustained health.

Investment in The Get Real Program will bring more to America in terms of quality of life and health for the many, not just for the elite. It will protect and preserve our leadership in energy and the environment. If not America, then who? China will continue with their efforts to advance in solar photovoltaic and other less effective solutions that alone do nothing for the longevity and sustainability of our environment. Their agenda is to become the world's superpower, not to preserve the health and life of America. Russia and China will surely take any opportunity to advance their own self-interests over the wellbeing of America and the rest of the world.

Estimates for The Get Real Program globally will cost approximately $1.2 trillion a year. America's share would only be $200 billion or about 30 cents per gallon of gas. While this seems like a large number, just think of many times more expensive climate packages that only serve to slow the rise of CO2 (and do nothing to lower the levels), but not to SAVE THE EARTH. It seems like a small price to pay when you think of it that way.

With every problem we highlight, The Get Real Program has potential solutions in hand. This can all be paid for by carbon sequestration funding to charitable companies willing to do the hard work of beginning major remineralization and creation of biochar (explained in later chapters), as well as changing agricultural and forest practices. Carbon sequestration efforts funded privately through charitable contributions would be part of this solution and is the most significant of solutions during our time – all paid for by those enjoying the benefits of this amazing planet. It is different from other proposals on carbon emissions that just try to discourage emissions by making them more expensive. This will fund the removal of more CO2 from the air than has been emitted; it will do so by solving other problems that most people agree are urgent but don't get the attention that climate change does. If voluntary and private solutions are not created, at some point the government will end up having to take over with wasteful and less effective strategies that are driven by special interests… and impacting our pocketbooks just the same.

Of course, our lens is beyond America, as we alone cannot make up for the practices and costs of other countries. Therefore, The Program would include incentive payments for other countries, like Brazil, to preserve and grow the rainforest instead of burning a huge amount of it every year. **This is a global issue and therefore, it will take a global solution.**

The Get Real Program can make huge strides, but we need investment to make this vision a reality. The Program solutions are not free; however, you will get what you pay for, as the saying goes. The Get Real Program addresses the intricate details of the carbon cycle, where real environmental opportunities exist. By reading and investing in real change as outlined in intricate detail in The Get Real Program and other gems provided throughout this book, you will be enlightened as to where real change is needed and why politicians and special interests are looking to solve for the wrong things in the wrong places.

"Real change cannot happen in a state of deliberation. The luxury of inaction has ceased. The solutions are at hand. Let's begin before our provisions end!"

– David Munson Jr.

CHAPTER 2

A TREASURE: EARTH'S CARBON CYCLE

The carbon cycle is essential to all life on this Earth. Life can exist without oxygen but not without carbon. There is a miraculous cycle of carbon from air, soil, plants, animals, oceans, and the Earth's crust. Some demonize carbon in the form of carbon dioxide (CO_2) with concerns over the environmental impacts of rising levels. Regardless of the beliefs you may or may not have around the topic of climate change, it is important to know that carbon is the key to advancing our way of life. There are approaches to make carbon work for us and not against us.

The solutions exist. Advances and innovation are critical for America and the rest of the world to respond to the fast-approaching threats to our way of life. It is not just advances that are needed, as there is also a place for creating a better future for us by utilizing what we currently have. Utilizing carbon at an ideal level can feed progress in regard to what are considered "climate change" challenges. The carbon cycle is not to be feared but fine-tuned to put the carbon where it will do good things – back in the soil and marine wetlands.

There are many climate activists out there purporting to create real change by eliminating entire industries. This may be very destructive thinking. Stepping back to old practices pre-innovation, rather than optimizing all solutions available to America, would be detrimental.

For instance, stopping the use of gasoline in cars completely or proposals trying to decarbonize electricity generation without increased use of other energies, like nuclear power, would be very misguided. It is important not to create new problems when trying to fix our current challenges. Balance is key. As an experienced farmer, the founder knows first-hand the value of preserving balance within nature to get the best output.

America should use everything available in advancing our natural elements to work for us. Ignoring any of our natural resources and elements will limit our ability to respond and threaten our sovereignty. Better yet, let's stack onto our advances and not take away any options. By having all resources available to us as we strategically advance, we will be more secure and have backup for the rainy days. Quite literally, we need backup sources to solar photovoltaic power, as rainy days can challenge the output of these energy systems. This will be discussed more in later chapters.

The Get Real Program can have the greatest impact when our environmental solutions are approached holistically, as the environment is a delicately structured system of organization requiring a multitude of advances at the same time. It is exciting to see there are very feasible solutions available to us today that can have huge short-term and long-term benefits. The solutions in The Get Real Program are advances that are simple, natural, reliable, and yet innovative. Let's learn more...

NOW, LET'S TALK CARBON

Carbon is a naturally occurring chemical element within the Earth's crust. It is carbon dioxide (CO_2) that has been polarized by climate activists, given that levels have been rising and there is concern that the increase is detrimental to our environment and its sustainability. The increased amount of CO_2 in the atmosphere has sparked much disagreement over both the direct and indirect impacts it is having.

As the levels of CO2 increase and insulate the Earth, there is a concern for increased temperatures, sometimes referred to as global warming. The truth is that CO_2 levels have fluctuated greatly over time, and there is greater impact from methane from natural seeps (much more than from oil and gas drilling activities) and increasing water vapor levels. Given that enormous quantities of carbon cycle in and out of the atmosphere, some climate skeptics maintain that the increase is of little consequence.

Some climate activists, for their part, ignore the fact that there is an active cycle of carbon going in and out of the atmosphere. It's not like you emit a pound of CO2 and it stays there forever. There's a continual movement that is larger than man's emissions by far. This is not to say that man's emissions should not also be considered, but it is important to have the right perspective.

CO2 levels actually go down in the northern hemisphere's summer season, because there's so much fertile land in that hemisphere. When the crops are growing, CO2 levels drop globally, because the crops are soaking up so much CO2. And then, once the crops stop growing in the fall, levels start rising again as decay takes the plant vegetation and reduces it or oxidizes it back into methane and CO2. It is an incredible experience to watch nature at work creating, recycling, nurturing, and sustaining life.

Arguments are also heated as to whether CO_2 levels in the atmosphere are too high at this point. The level goes up and down, and one of the helpful actions we can take to impact the levels of CO2 in the environment is to keep growing things in the soil. It is especially important in regions where crops can grow year around, like Texas, instead of farmers leaving the soil bare. Continual crop growth and other efficiencies in farming practices can provide even more benefits.

If there is a level of CO2 that is toxic that will make life really harsh, slowing the increase doesn't save things. You have to actually start to stabilize the level or reduce it. To do that, you need to focus on the solutions of The Get Real Program, like increasing carbon sequestration, as well as being sensible about reducing carbon emissions where you can. The sequestration potential is high enough that we can continue to use the level of oil and gas that we use now.

You would think that with such an interest in CO2, there would be major efforts to track carbon emissions on a global basis. Instead, much of the attention has been focused on the easy-to-track fossil fuel emissions as the quantities used are recorded and easily available. The carbon emissions from the soil, a forest fire, or a volcano are harder to estimate and nearly impossible to pinpoint. Yet, there is so much more potential for impact by working with nature to sequester carbon than focusing on man's emissions.

An unhealthy carbon cycle caused by a demineralized world is a real threat to long-term survival. The Get Real Program knows that it does not matter what side of the climate change debate you are on, as the practices that can help lower CO2 are needed to advance America on other environmental fronts too. The changes realized by remineralization and carbon sequestration are a byproduct of needed changes on multiple fronts, all without ending comfortable American lifestyles.

Part of our American lifestyle and rights include our freedom to eat what we desire. The climate activists are very wrong to indict animal husbandry as an undesirable activity that increases greenhouse gases and degrades the environment. What is needed is a different way of raising beef, in particular, moving away from grain-fed beef and toward grass-fed beef. Managed and controlled grazing that works in harmony with the grass can sequester many tons of carbon per acre per year, whereas clean tillage grain farming is a net emitter of carbon. There is more detail on this important subject in Chapter 6.

The Get Real Program understands how important it is to be comprehensive and focused on the long-term impacts, while addressing short-term practicalities. The issues within the environment were not created overnight nor will the solutions have impact overnight. The Get Real Program focuses on holistic health for this Earth, regardless of where you stand on what the issues of carbon are or are not.

Now, let's learn more about the carbon cycle.

THE CARBON CYCLE

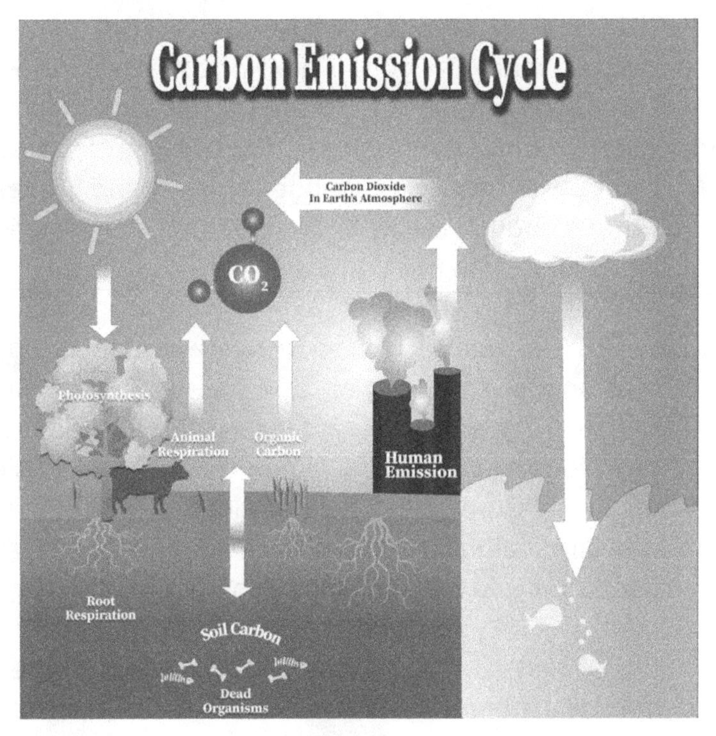

Figure 1

Many who write on CO2 make lots of graphs and numbers to describe the carbon cycle, but the truth is that it is impossible to know where all the carbon is in the cycle. However, we can know pretty closely how much CO_2 man emits by combustion of fossil fuel. Even that number can be off, though, as some fuel doesn't burn into CO_2, and some turns to soot or carbon particles.

We can use the atmospheric CO_2 level at the top of a mountain in Hawaii as a benchmark for the air, but in truth, the level varies depending on plant life and other factors. The CO_2 level in a crowded bar with poor ventilation may even induce a bit of oxygen deprivation due to all the carbon dioxide-rich breath, while a corn field may approach zero CO_2 in peak growing time. We do know that atmospheric CO2 doesn't go up as much as man's emissions.

The Earth's carbon cycle is incredibly complex and a true wonder. The cycle is a miracle of movement of carbon from the atmosphere to the Earth in an automated and natural flow that goes unseen by the naked eye. Currently, the cycle is skewing towards a net release of carbon at a large rate because there is no incentive to create opportunities to remove it from the air. The stores of carbon in the cycle are generally dynamic and change with time and the season. The amount of carbon in the air at any given time is a fraction of the annual flow of carbon through the cycle.

Man is part of the carbon cycle in so many ways. Climate activists tend to focus on man's burning of fossil fuels alone, lumping all fossil fuels into a group without regard to the significant differences among the various hydrocarbons. *If* man quit burning fossil fuel immediately, greenhouse gases would continue to rise due to man's practices, as well as natural processes. Not very much of man's fossil fuel emissions remain in the atmosphere long term.

Combustion of fossil fuel is a significant source of higher CO_2 levels in the atmosphere, but it is far from the only source. It is, in fact, dwarfed by the net carbon emission from the natural world due to man's activities (such as deforestation) and the documented demineralization of the soil as well as natural seeps from tundra and other sources (methane ultimately degrades to form CO2). There are more microscopic life forms in a square foot of healthy soil than there are people on Earth today, and the soil's role in the carbon cycle is more significant than man's burning of fossil fuels.

The big-picture carbon cycle chart above (see Figure 1) does not go into detail about the soil-carbon interaction. It just indicates that carbon goes both into and out of the soil due to a variety of processes. As a successful farmer and rancher for many decades, the Founder can tell you that healthy soil could remove a lot of carbon both directly from the atmosphere and indirectly, primarily through the nurturing of plants and animals.

We have to increase the amount of carbon removed from the air by the soil, by the plants, and both on land and in the oceans, to have an impact on the health of our environment. Slowing emissions is not enough at this point. There is more to be done to create real change. Carbon must be both sequestered and managed in regard to emissions. This will take a comprehensive approach, but it CAN be done within The Get Real Program.

CARBON DIOXIDE [CO2] LEVELS

It is important to get our CO2 levels in check and working for us. By looking at the entire carbon cycle process, The Program demonstrates opportunities to advance. The future does not have to be dire, but a more prosperous one that works with what we have and creates more. The mismanagement of our resources is at a tipping point, but there are positive solutions.

Exposure as a farmer and rancher to real issues in the field has given the Founder a unique view into solutions that some scientists cannot even imagine in their limited experiments. Lifetime field experience is crucial in knowing and understanding the process in order for us to work within it and not against it. There is a pathway to a more prosperous future, even with the CO2 levels where they are today. We can work together to make America carbon-negative with our innovative creations and a few actions by the masses.

Yes, the levels are concerning. however, the solution is to work with the problem, sequestering the carbon to put it to better use and therefore alleviate the harm it is doing to the environment. There is debate as to whether and to what degree the increased CO2 levels are creating problems for the environment, but either way, why not optimize its use and put it to work for us? The answer is in the soil and other natural solutions to create change over time to support advances.

THE SOIL SOLUTION

In farming, we have seen the deterioration of our soil as a result of mismanagement. A major reason CO_2 levels have risen is the continuing expansion of the world's deserts, reducing the amount of carbon stored in the soil and plants. Not only are we losing our fertile land by the day, but the downstream impact of the deterioration of our minerals in the soil is causing tremendous health concerns based upon the foods we are eating. Something has to be done. Our bodies are not made to replace all our own nutrients. We have to get the minerals from somewhere, and that place is Mother Earth.

Many climate activists are city dwellers with little real knowledge of the potential of the soil to store carbon if managed properly, as well as its ability to release CO_2 to the air if mismanaged, which is what is occurring today. The answer does not lie in just one solution, but in a comprehensive program that can impact the entire cycle. We hear so much about the rising temperatures linked to the mention of rising CO_2 levels in our environment. As a farmer, the Founder has seen the changes in our soils, waterways, and environment – changes that he wishes he could forget or were not reality. We can tell you there is so much more to worry about than rising CO_2 levels, and the effects have compounded over time to create new problems.

Yes, lowering overall CO_2 levels by increasing plant and soil life to absorb carbon that has been released into the environment might increase average temperatures, but averages don't really matter as much as extremes. The average daily temperature in the desert might be 75°F, which seems very pleasant, but it is the high of up to 130° during the day and lows near freezing at night that happen without the temperature-damping effect of water vapor. There is not a carbon cycle in the desert, as all of the carbon that should be present in healthy soil has been oxidized away in the dry desert soil and atmospheric CO_2 is not being cycled through vegetation and into the soil. This is just the beginning.

Curiously, the growing deserts actually cool the Earth, as they reflect much solar radiation back to space. They are generally cloudless and have little water vapor to trap heat. Converting a large desert to forest would not cool the Earth. Converting a desert to forest or grassland would sharply lower CO_2 levels but would increase average temperature due to increasing water vapor. This is positive as we need a more moderate climate. In fact, over millennia, the global warm periods correlate highly with increased prosperity and strong economic growth. This is why The Get Real Program advocates for increasing the carbon cycle storage away from the atmosphere by dramatically improving soil and plant life.

Oftentimes, environmentalists operate from a mindset of not fighting nature. We need to actively fight harmful natural processes and intervene to fight desertification in strong ways. Many of man's unnatural ways got us where we are today. We must fight nature with nature. Altering the carbon cycle by removing CO2 from the atmosphere and simultaneously improving soil fertility creates nearly permanent change that is beneficial in many ways. The Get Real Program will put CO2 to work for us, so that we can rest easily knowing that we are nurturing our environment.

HOW TO MAKE CARBON WORK FOR US AND NOT AGAINST US

There are many solutions laid out in The Get Real Program that work within the carbon cycle to advance a healthy environment. The Get Real Program advocates for modifying the natural processes of the carbon cycle so that the CO2 level in the air actually goes down in a beneficial way, not trying to come up with ways to extract CO_2 from the air by technical processes.

Just let nature do the work while also improving the soil and oceans! The carbon cycle is interdependent. Far more carbon exists in the soil than in the air, but it is a balance that has been skewed by man reducing the soil's carbon content through bad practices, which has led the level in the air to be higher than some think it should be. Simple, doable changes can change the atmospheric level even while we burn fossil fuel. The gains from having living, mineral-rich soil will be numerous, ushering in an age of prosperity and ample food for man and nature.

You can't lower levels of CO2 without increasing sequestration in the cycle. Current proposals that focus on cutting fossil fuel emissions as a long-term goal don't do anything but slow the rise. The Program proposes steps that can actually lower CO2 levels through increased biological sequestration with a global effort.

The cycle is worldwide, but man can alter it by increasing photosynthesis and lowering atmospheric CO2 levels dramatically while still using fossil fuel. We can see real change with improved agricultural practices, remineralization, large-scale use of biochar, and rejuvenation of marine wetlands. It is a total win-win-win with only a few industries suffering, and those industries are very unsustainable anyway. The solutions exist!

Some key areas where we can create real change include:

Soil

Soil is one of the Earth's most prized resources, yet it is too often overlooked for its potential. The benefits to be gained from the richness of a healthy soil cannot be underestimated in the fight for a healthy environment. There are solutions that will keep our land fertile year-round, which can fundamentally change the carbon cycle. This can build soil carbon in a sustained way, instead of the current annual ebb and flow. Remineralization and carbon sequestration within the soil are among the most effective ways to create drastic change quickly.

One thing that can't be overlooked is that carbon is an essential nutrient for plant life of all types and is often the limiting nutrient. Some of the gains in farming are due to extra CO_2. It is said that during peak growing time, within rows of corn, the CO_2 level drops to near zero because the corn plants are consuming the CO_2 to grow at a fast rate.

Researchers rarely deal with truly living soil, and mixing and even sterilizing occurs with test soils. Only an experienced farmer can truly see the keys to getting the most out of our land. If a square mile of farmland was devoted to prairie grass on the portion that can be serviced by a circular pivot irrigation and harvesting mechanism, the unirrigated corners could be used for cattle sheds and algae-growing ponds. We can see dramatic changes by fully utilizing our land. Let's work together to see a comeback of our rich soil.

Deserts

A large part of the land is nearly sterile and not part of the carbon cycle – the world's deserts, which are continuing to grow in size. In some years, over a million acres fall into desert conditions. We should have a goal of reversing that with a million acres a year going back to vegetation.

Desert soil with no organic matter and no soil life oftentimes can't absorb water, and when it does rain, the water runs off, carrying small clay particles with it. The process of desertification releases millions of tons of carbon back into the air as the environment degrades from fertile to sterile. Like maggots eating a carcass, the decomposing organisms consume all the carbon stored in the soil and plants as the land transitions. In desert areas, grass has to be replanted and nurtured. Restoring desert and preventing desertification from occurring will take money and effort, but it can be done, and the results are great for the world at every level.

Regenerative Grazing

The impact of cows on our environment has been a very popular topic in the media. We dare to say that cows are part of the solution, whether you feel they are a problem or not. Using the knowledge of the past and incorporating it into the future will be key to solving the health and nutrition requirements on this planet.

Regenerative grazing can restore the land from near-desert to lush grassland, which can have a positive impact on the carbon cycle and the health of our environment overall. The key to regenerative grazing is to concentrate the animals for short-term grazing followed by a period of re-growth. Thus, a beneficial step is not eating less beef but changing the way we raise it, so that carbon is sequestered in abundance. In Allan Savory's book *Holistic Management*, he postulates that to restore land to grassland, what is needed is not an end to grazing but a change to the right kind of grazing that actually helps the grass thrive. The great thing about this method is that it will produce more food for a hungry world.[1]

Old ways of doing things often go on long after a new and better way is developed. The current way of raising cattle needs lots of changes to make it carbon sequestering instead of carbon emitting. Many climate activists are ardent in demanding that we eat no meat to "save the planet," but actually what is needed is more animals raised holistically. Grassland under holistic management is a carbon

1 Savory, Allan, with Butterfield, Jody, ***Holistic Management: Third Edition: A Commonsense Revolution to Restore Our Environment***, Washington, DC: Island Press, 2016.

sequestering wonder, with the carbon sequestered far exceeding the carbon emissions of properly pastured cattle thereon.

Grassland that is not grazed at all declines just as badly as grassland that is improperly grazed using traditional methods. The key to healthy grassland is to simulate the activity of herd ruminants under pack predator pressure, which is how grassland evolved. Allan Savory says if you manage cattle correctly, they're actually carbon-negative, because they're encouraging the grass to grow, and their manure fertilizes the grasses and builds up organic matter.

Biochar

Biochar, a partially oxidized organic plant remnant which retains most of the carbon but not other more volatile compounds, can be created from undesirable existing vegetation. Incorporating the beloved and life-initiating biochar into the soil will help hold water and nutrients, in addition to permanently removing carbon from the carbon cycle. It also creates a sponge for water in the soil that will nurture soil life and return dead soil to life.

Biochar does not have a nutritional value directly tied to it for plants; however, The Get Real Program has ways of priming this resource for advancement of soil life and reclaiming deserts in favor of the health of the Earth's environment, including water retention and micro-biome habitat generation.

Forests

Trees can make a huge impact on the health of our environment. Forests are a huge factor in the climate equation, and the die-off of the trees or massive fires hurts the water cycle and the absorption of water by the soil.

As you continue to read, you will see how a comprehensive view of The Program is necessary, as all of Earth's natural systems work together, and not independently, to either solve or create problems for the environment. Forestry is an area where we want to see increase, but there must be a strategic approach to get the balance and results needed for carbon sequestration, proper forest management, and increasing healthy biodiversity.

Oceans and Wetlands

The oceans occupy most of the Earth's surface and are major players in the carbon cycle. Therefore, this is an area requiring much investment and continued research to find all of the ways to create impact. Ocean restoration is a massive feat, but it is imperative to learn more, as it holds the vast majority of Earth's CO2 (actually as dissociated into components such as hydrogen ions and carbonic acid). There are some clear answers today, like restoration of marine wetlands.

Marine wetlands are huge sinks of carbon that are being wiped out by man's activities of dredging and filling. Restoration of marine wetlands and shallow water areas is one part of the holistic solution of carbon sequestration laid out in The Program. Researchers and scientists are somewhat divided on the desirability of adding nutrients to the oceans. The Get Real Program understands the complexity and knows that a bold action and new methods are needed for sustainability. Most of the ocean is very low in life, so there is a lot of opportunity for added carbon sequestration in the ocean. Also, areas at river mouths have become dead zones for many reasons and The Program can help address this issue.

Smart Energy (Realistic Renewables)

In the long term, we need to make greatly increased amounts of renewable biogas from waste of all types. As will be discussed later, natural gas is the ideal backup for wind and solar photovoltaic, which are, by nature, variable and intermittent. While this book advocates for funding natural ways to reduce CO_2 levels, it is desirable to rapidly phase out coal use, and the only practical way to do that is with substituting natural gas, which is cleaner and more efficient.

Nutrient Recycling

No plan of action on the environment will be highly effective in the long term without changing the way we fertilize crops and increasing the soil's ability to hold nutrients. Soil mismanagement and degradation are more threatening individually than the rising CO2 levels. Once the nutrients are gone due to a lack of managing the process for the long term, then so is life on Earth.

Rock Dust

Without rock minerals, life cannot exist. Improving the soil by adding relatively water-soluble rock dust, highly concentrated with a nutrient found to be deficient on a particular farm, is a good idea in the short term. However, putting on nearly insoluble rock dust containing a wide variety of nutrients has many positive, almost miraculous, effects as well.

Carbon Sequestration Strategies

Carbon sequestration will have the largest impact. It is a fundamental part of The Get Real Program that can reduce atmospheric CO_2 levels even while we continue to burn oil and gas. Again, balance is key. The Program offers creative improvements to support better environmental health, but there is also a need to adjust our current coal mining and other wasteful practices to avoid progression of harmful effects on our environment. The Get Real Program can alleviate burdens that increased CO2 levels create on the environment, but without dramatically changing our way of life.

THE GET REAL PROGRAM BELIEVES IN USING NATURE FOR CHANGE

While there are many solutions for working within nature, changes to man's practices also have to be made. Critical to these efforts, The Program feels it is imperative to sharply reduce coal burning for many reasons, namely because coal is far and away the highest carbon-emitting fossil fuel. Sadly, many organizations focus entirely on trying to minimize man's emissions while ignoring the fact that over half of the fossil fuel emissions are already sequestered. There are a number of people who actually want to see natural gas use go down, keeping coal-fired power plants in operation. They think that it is good for the environment long term, when it actually makes it worse. Without natural gas power plants, you can't efficiently back up the intermittent power sources, like solar photovoltaic or wind. This is why it is so important to get real with the facts and share The Get Real Program broadly.

Coal is carbon-rich, so The Program can see a quick "bang for our buck" by phasing out coal. The impact is greater than cutting back on oil and natural gas use. By using coal, carbon emissions would remain high for the foreseeable future if banning new natural gas power plants. There is so much more opportunity to cut carbon emissions in half, or more, by phasing out the coal plants. So much can be done with the right approach and a commitment to making the world a better place.

The key opportunity with the carbon cycle is to make it work for us – lowering the level of atmospheric CO_2 by increasing the amount stored in the soil, plants, and ocean while reducing the release of CO_2 by dying trees and undue tillage of the soil. It is vital to make significant changes to the way we do things on land and sea. Happily, proposed changes such as adding cover crops, soil remineralization, extensive application of biochar, regenerative agriculture, restoration of wetlands, and others are net winners and very doable.

The world has urbanized at the expense of mismanaging the countryside, putting all life at risk as the environment degrades and vital nonrenewable resources are squandered. Chief among those resources is the soil, which is being eroded and degraded into oblivion. Change is needed right away because every day that we wait there are more deserts created, and more topsoil washes into the sea. The rich topsoil of the Midwest was created from nothing but rock dust, water, and photosynthesis after the last ice age. It is an amazing thing that organisms can turn rock into soil. All we have to do is let them do it by giving them rock dust to eat.

With the right management methods and practices, we can easily reduce the CO_2 level; it should be a matter of scientific debate as to the optimal level to maintain. We can solve the supposed climate crisis by fixing other very real and serious problems affordably and much more quickly than the extremely long-term programs that have been proposed. Generally, when politicians propose something unworkable, they delay its full implementation until far into the future, long after they retire. They make promises today that are paid for in the distant future, or the consequences come due.

Let's GET REAL and solve the CO2 issue measurably in a matter of a couple of decades by changing the carbon cycle from increased emissions to major sequestration. This will benefit man with improved soil productivity in the near future, not many decades from now. With billions of tons of soil lost and the addition of a million new acres of desert each year, we can't afford to wait. See www.getrealalliance.org for more information and to support the solution.

Let's GET REAL on Carbon and Make an Impact Today. The Clock Is Ticking!

CHAPTER 3

THE RICHES: ROCK DUST AND REMINERALIZATION OF SOIL

"Soil is apolitical. Whether you're liberal or conservative, Democrat or Republican, urban or rural, ride in the bicycle lane or drive a pickup truck, the well-being of the earth matters to everyone. Soil remineralization is simple, intuitively learned, and applicable at the community level. What is required is the planet's most abundant resource. Rocks! We can move from an era of economics based on scarcity to one of abundance through the power of rock dust."

- Carter Haydu, Energy Journalist and Remineralize the Earth Volunteer Writer [2]

Quite simply, finely ground rock dust from many available sources has a broad spectrum of minerals and trace elements that are essential for life on Earth. The lack of minerals in the soil is the real existential threat to man that far exceeds any threat from CO2 by an order of magnitude. Without remineralization with rock dust to

---

[2] **Remineralize the Earth** (RTE) is a 501(c)(3) non-profit organization based in Northampton, Massachusetts, and founded in 1995 by Joanna Campe. https://www.remineralize.org

create a better environment and clean up our food supply chain, we will not live in a world of abundance.

WITHOUT ROCK DUST...

Without efforts to remineralize our soil back to health by embracing solutions that include rock dust, we will not have healthy soil.

Without healthy soil full of minerals, health in our plants will diminish.

Plants without adequate nutrition become hosts for disease and insects. The sick plants will attract microbes and various parasitic organisms. Plants that are not healthy cannot maximize their potential for carbon sequestration. Unhealthy plants serve an unhealthy food continuum towards an unhealthy existence for all.

Without healthy plants, people and animals can get sick from what they're eating and the lack of nutrients from their food.

Without plants optimized in their role for the health in our environment, we may all suffer.

Without a healthy environment, we will ultimately die off.

It is all a cycle that needs to be healthy in order to have healthy outputs, which translates to a healthy environment and healthy lives. We're in a dark situation with our soil and must take action now! The future can be full of life with just a little sprinkle of rock dust.

LIFE IN THE LIFELINE... ROCKS AND ROCK DUST

Without rock minerals, life cannot exist. Not all rocks are created equal, though. They vary widely in mineral composition, with many rocks largely lacking in vital nutrients required for a rich soil and nourished plant life. Some of the best soils originate from silicate rock, which originates deep within the Earth, and with volcanic activity compose part of lava hardening at or near the surface. This process is continuous, constantly establishing new surface rock from molten rock stores from below the crust.

Weathering by wind and water breaks the silicate into small pieces, which microorganisms then attack, extracting nutrients. These microorganisms leave behind clay particles so small that they flow like water. Gritless and smooth, clay has a cation-exchange capacity, meaning it can hold nutrient ions. Unfortunately, because clay particles are so small, they easily erode when combined with water to form mud.

Organisms break down the rocks to extract nutrients and make other compounds, often using CO2 through photosynthesis. In fact, the world's most fertile soils exist where nature has already broken-down rock matter for the life forms that feed on it - where glaciers have crushed rock into powder and where volcanoes have scattered ash and rock particles. Compared with the lush, richer soils of mountainous volcanic islands, islands formed from limestone have lower soil fertility.

Rock dust can unlock more opportunities to advance our lives by fueling the earth to support plants in their use of CO2. The goal of The Get Real Program is to increase photosynthesis and carbon sequestration so that the world has a falling CO2 level, even while still burning oil, natural gas, and some coal where necessary. All of this can be accomplished while also protecting our overall energy sources, which will secure America's sovereignty. Nature has provided solutions for us over time, but man's practices in not nurturing our natural resources can only take us so far. It is time to reinvest in our natural resources and learn from what we have seen over history to help catapult us into a better and healthier tomorrow. We only have to look at what is right in front of us to create change.

Let's explore more about our precious rocks...

Basalt

Earth has an abundance of rocks, but the best type of rock for remineralizing the earth is basalt. Basalt has a unique, wide array of nutrients, and it's shown to deliver significant improvement agriculturally. It is one of the most common rocks on Earth, but it's not distributed uniformly across the Earth. It's a volcanic rock, and so there's an abundance of it in areas where there's been volcanic activity. In the event that basalt is not available nearby, rail transport and distribution by specialized trucks can often be efficient ways of moving it.

The Get Real Program has innovative solutions for transport and other needs of getting the rock dust into just the right particle size and delivered to the right spot for the greatest penetration that leads to the best outcomes. See more on this topic later in this chapter in the section *Technology to the Rescue.*

There is more than one rock type that can create real impact, but if only one is used, basalt is the best.

Limestone and Dolomite

Limestone and dolomite can also be great for our soils, because they are concentrated in calcium and magnesium. Life is about abundance. A combination of nutrients in our rock provides the greatest foundation for Earth's ecosystem. If the soil is deficient in magnesium, it needs dolomite. For calcium deficiencies, limestone is the answer.

Limestone is precipitated calcium carbonate from ocean life, oftentimes containing fossils originating from deposit locations on the seafloor. It has very few other elements in it. Limestone is soft, easily penetrated, and leaches away with rainwater. In fact, many caves form when rain carves away the limestone, creating caverns. FYI: antacid tablets are ground up limestone or dolomite that rapidly raise hydrogen (pH) in the stomach and are used to treat acid reflux.

One might guess, therefore, that crushed limestone on soil works quickly, while the elements in basalt (which are not water-soluble) would be unavailable. Scientists who work with dead or nearly dead soil promote water-soluble nutrients because the plant roots can take them into the plant directly. However, in the medium and longer terms, non-soluble minerals are key as well. Limestone-based islands would greatly benefit from soil remineralization using basalt rock dust. In practice, basalt produces significant results, especially on demineralized soil. *The evidence is compelling!* The Get Real Program looks to comprehensively use natural resources, including primarily basalt and also limestone, to remineralize the earth for the greatest potential in our crops and the environment.

Shale and Sandstone

Depending on where it is deposited, the hardened mud known as shale can offer in some cases very little in terms of minerals and in others some good mineral content based on the cation-exchange effect discussed previously. Shale rock is clay that has accumulated in the sea, eventually covered by some other material until enough pressure is applied to make it solid. Likewise, sandstone usually contains very few earth elements with the exception of silicon, except in cases of mixed sand and shale depositions or other special cases.

THE GLACIER CONNECTION TO OUR SOILS

Glaciers have created a pathway to a better future through enrichment of our soils with glacial rock dust. It is a long, slow process by which nature creates fertile, deep soil. For glacial rock dust, this process began thousands of years ago during the Last Glacial Maximum, as vast ice sheets at least a half-mile thick cut deep into the bedrock, erasing what topsoil preceded it. We can learn from the glacial past to solve for the present deficiencies in our soil today.

In *The Survival of Civilization*, the seminal book on remineralization that launched a grassroots movement in the 1980s, John Hamaker presented a study that showed his farm remineralized with glacial gravel dust. The soil yielded 65 bushels of corn per acre with no irrigation, compared to 25 bushels per acre grown on nearby farms using conventional fertilizers. His corn had a 28 percent increase in protein, 47 percent increase in calcium, 57 percent increase in phosphorus, 60 percent increase in magnesium, and a 90 percent increase in potassium compared with the control.[3]

Imagine what it would mean to increase the nutrient density of corn and other produce to this degree on poor soils in Africa and elsewhere. The right rock dust can make the world a much better place. This book stands on the work of those researchers and farmers who understand that we can lower atmospheric CO_2 levels while growing more abundant, healthier food.

We can reproduce what the glaciers have done, grinding up rock into dust and applying it to land and sea. In particular, the application of basalt rock dust is a proven effective soil remineralizer, and many millions of tons of it are already available worldwide as a byproduct

[3] Hamaker, John D., Weaver, Donald A., ***The Survival of Civilization Depends Upon Our Solving Three Problems: Carbon Dioxide, Investment Money and Population - Selected Papers of John D. Hamaker,*** Hamaker-Weaver Publishers, First edition, June 1, 1982.

of the resource economy. Today, the utilization of rock dust for remineralization of the soil at the level needed to stave off our increasing soil depletion will only happen with human intervention and investment, both yours and mine!

THE MIRACLE OF SOIL LIFE REVITALIZED THROUGH REMINERALIZATION

Investment in our soil through remineralization can bring about incredibly drastic advances to our environment through reductions in carbon from increased carbon sequestration into the soil, as well as a healthier planet overall. Humans and animals alike will feel the difference as our bodies renew from the increased nutrients in the food supply chain. It has been so long since there was a widespread healthy food supply that most people have a limited idea of what it would be like to have really nutritious, mineral-rich foods, rather than just surviving with a basic and depleted nutritional food supply.

Soil biology can make insoluble elements available to the plant roots, extracting elements from the rock dust. Rock dust is literally food for soil microorganisms. Well-fed microorganisms nutritionally benefit plants with which they share an ecosystem. Each year, farmers worldwide dose cropland with concentrated phosphorus and potassium, which largely leaches away to the sea or is chemically locked up in soil due to poor microbial action in what has become nearly dead soil. An advantage to broad-spectrum insoluble rock powders is that they do not leach away as do concentrated water-soluble plant nutrients (i.e., conventional fertilizers).

It is still relatively unknown as to the optimal number of elements for plant growth for producing the best food. Through analysis of plant materials, though, we can see some of the nutrients present and have a pretty good understanding of them, especially higher-concentration nutrients that can be studied for deficiencies. When it comes to the optimal number of elements in soil to promote plant growth, there is a Goldilocks Zone, with too much or too little of any given element negatively impacting optimal growth and nutrition. Water solubility and reactivity of elements vary as well, and the advantage with utilizing basalt rock dust for remineralization is that the soil biology will extract nutrients from basalt and this occurs before any "leftovers" become water soluble.

Measuring the hydrogen (pH) of soil or water is easy, and for many years, scientists considered it the most relevant thing to measure in soil. However, William Albrecht (1888-1974), who was chairman of the Department of Soils at the University of Missouri, determined that, although near-neutral pH is preferable, it is also important that basic elements are in the right proportion as well for optimal soil chemistry.

Soil becomes increasingly acidic as rock nutrients leach out with rainfall. Soil that is either too acidic or too basic does not allow plants to grow well. As important as biology might be for healthy soil, chemistry is also key. For example, some elements are unavailable at either low or high pH levels due to chemical reactivity. Soil life itself cannot survive in acidic (low pH) soil, as that means there are no available rock nutrients. The so-called cation-exchange capacity of soil measures soil's ability to hold positive nutrient ions such as calcium and magnesium to clay particles and humus.

Ideally, soil cations are attached to soil nutrient ions for plant availability. Albrecht was a pioneer in determining the ideal amounts of nutrients in the soil, observing that a mix of about 68 percent calcium ions and 12 percent magnesium (with lesser amounts of phosphate, potassium, and other) allows for optimal growth of many types of crops. Sadly, most of the world's soils are so terribly demineralized that they simply do not have anywhere near that amount.

There has been a massive decrease in nutrient density in our foods over the decades; however, soil remineralization can rapidly increase the nutrient profile of foods to a more natural, original, nutrient-dense state. While it would be ideal to remineralize all the world's soils for optimum nutrient balance, any improvement helps. The company Rock Dust Local provides more practical information on rock dust and soil amendments. Gardeners and farmers can visit www.rockdustlocal.com for rock dust, biochar, and other important products to sustain and restore soil.

Many farmers just leave their soils acidic, sometimes making things even worse by applying lots of excess nitrogen. Many regions of the world lack nearby deposits of limestone or dolomite, and it can be very expensive to bring in the rock dust. In these regions, sea minerals are an alternative.

Highly diluted sea water and concentrated sprays are another very practical way to remineralize soils. For that reason, financially strapped farmers do not need to cut corners by applying the wrong nutrients, especially when doing so reduces quality. There are many sustainable and affordable options to consider in the transformation from industrial to regenerative agriculture.

ROCK DUST TO THE RESCUE THROUGH REMINERALIZATION

Applying relatively water-soluble rock dust, highly concentrated in a missing nutrient, is a good idea, perhaps obviously, at least in the short term. However, putting on less soluble rock dust containing a wide variety of nutrients has many positive—almost miraculous—effects. Many tests on demineralized earth show positive improvement in both plant growth and quality when basalt rock dust is added.

Remineralizing the earth is essential not only for sustainable agriculture and promoting abundant life on this planet, but it is key to combating climate change as well. A research team within the Leverhulme Centre for Climate Change Mitigation led a study that incorporated finely crushed basalt rock dust into agricultural soils for sorghum. Their findings showed that this can greatly improve crop yields and can result in storing impressive amounts of carbon from the atmosphere in the remineralized soil. *"Enhanced rock weathering"* is the term utilized by climate and soil scientists like those at the Leverhulme Center to refer to the remineralization of soils with an emphasis on capturing carbon to stabilize the climate.

Soils are part of our planet's carbon cycle, which limits the amount of CO2 in the air and regulates the climate. CO2 in the soil reacts with rock dust to form bicarbonate and calcium carbonate. Bicarbonate is soluble in water and will eventually make its way into the oceans where it counteracts ocean acidification. Calcium carbonate is a solid mineral that can store carbon for millions of years. Eighty percent of all the carbon on Earth is stored in rocks as calcium carbonate. Rock dust remineralization of soils takes advantage of the most abundant carbon sink on the planet to passively remove CO2 from the atmosphere while also improving soil fertility!

What Can Be Done?

When it comes to feeding the world's population and keeping climate change in check, few proposed approaches are as attractive as soil remineralization. Introducing the right minerals into soils provides nutrients for plants and elements that react with CO_2, pulling it out of the air and into the ground, where it could potentially be stored for centuries. In the study, compared to plants with untreated soil, soils treated with basalt rock dust revealed a 21 percent crop increase. The sorghum absorbed 26 percent more silicon, increasing the plants' ability to fight off fungal pathogens, resist pests, and increase their stem strength. Basalt rock dust also increased carbon capture fourfold!

In the long term, remineralized plants are naturally more resistant to insect predation, but rock dust can be used in the short term for an insect infestation. Insects have a naturally waxy protective covering, which normally keeps them from drying out. When sprayed directly on insects, the rock dust interferes with this covering, gets into the segments, and disables them. During an infestation, this can restore insect balance.

The Leverhulme Centre for Climate Change Mitigation's study demonstrated that *enhanced rock weathering* using basalt rock dust on sorghum showed promising results in increasing crop yields, establishing insect balance, and capturing carbon. If we are to restore soil health and enable its carbon sequestering capabilities, we must know what makes healthy soil healthy and be able to rejuvenate unhealthy soils. We must know what is in a given soil to properly address the deficiencies.

Testing Is Key

With good soil tests, farmers can add the ideal amounts of calcium and magnesium, bringing the soil exchange capacity into balance. Sadly, though, many farmers do not test the nutrient balance of their soils, as do hardly any ranchers. Where tests do occur, some of the laboratories use timesaving tests that fail to accurately show what is available.

The Get Real Program puts an emphasis on soil testing on an annual basis in order to know where there are deficiencies in the health of the soil. To get the most of our soil, a critical ratio of nutrients must be maintained between calcium and magnesium. Testing the soil can show where nutrients are leaking out, so that farmers can remineralize with rock dust to maximize their crop outputs and carbon sequestration.

Size Matters with Rock Dust

It's desirable to turn the rock dust into an extremely fine powder similar to the consistency of flour, so that the microbes can access all the nutrients in it. A microorganism can only eat the surface of the rock. They can't bore inside the rock. Therefore, if a particle is an eighth of an inch in diameter, the microbes can only eat a thousandth of an inch or so. The ideal particle size is about a thousandth of an inch, so that the microbes can essentially access all the nutrients in a pound of rock.

If you put a big one-pound rock on the soil, the nutrients that are available to the microbes is only maybe a thousandth or ten-thousandth of the mass of the rock, because all they are doing is eating the outer surface of the rock. This is why grinding the rock to a fine powder and pelletizing it will keep it from blowing away in the wind. The fine dust stays suspended in air almost indefinitely. Dust from the Sahara Desert travels all the way to the Amazon Rainforest before rainfall washes it out of the air, and thankfully into the soil.

Small amounts of basalt rock dust are harmless, and if applied in an inhalation-free format, pose no hazard to animals or people. That being said, it is not good to inhale silicate rock dust. Basalt is a silicate rock, so you don't want to be in a big dust cloud breathing it. That could cause silicosis of the lungs. A pelletized version or a dust dispensed in water is best, where the rock dust goes onto the soil instead of into the air.

Read further to learn of new and exciting technologies to support rock dust and remineralization.

Compost Is Key

Life is a mystery. Why does it happen? How did it start? How small can something be and still be considered alive? Why does basalt dust cause such significant soil improvement? The real proof of rock dust's effectiveness comes with improved performance. In some areas, limestone and dolomite are badly needed but unavailable. It is true that they leach away in high rainfall areas, especially with acidic soil, so basalt (which does not leach) may be a better choice, even though soil chemistry *theory* suggests it is not that great. Applying basalt rock dust does little without active soil microbial life to digest it and create new, valuable compounds.

Generally, soil enhancement studies have much better results when they include inoculation. One of the great ways to revitalize soil is with well-made compost. Quite simply, a good compost pile is amazing. In Brazil, for example, studies show remineralized "sweet" cacti can far surpass the normal productivity for the species, especially when treated with rock dust plus compost. See www.remineralize.org for more information on these studies. In the study, within the first year, the cacti grown in remineralized and composted soil had 3.9 times the amount of growth of the control group. The study showed a fourfold increase in the presence of calcium and a twofold increase in magnesium, which greatly improved the soil pH.

Organic matter will decompose if exposed to air. The key to making really good compost is keeping it aerated so it does not become anaerobic (lacking oxygen), ideally inoculating it with organisms that turn the mixture into humus, a mysterious compound that exists in healthy soil. Formed by the decomposition of plant material, humus gives off a wonderful earthy aroma, a sign that it can easily degrade into gases and particulates. In healthy soil, humus can provide long-term carbon sequestration that benefits soil and plants greatly.

Humus is largely a carbohydrate material, meaning it is composed of carbon and hydrogen with few rock elements in its structure. It serves as a holding compound for ions of critical nutrient elements in healthy soil. Plant roots are both intakes and outlets for nutrient and "waste" compounds. They take in desired complex compounds, as well as ions of needed elements, transforming them into plant matter, and they return "waste"/respiration compounds and excess nutrients and water to the soil

There is more that we can gain from our investment in rock dust that can compound into great benefits for all, including some benefits from paramagnetism.

A mystery within the paramagnetic advantage

The role magnetism plays in the world of soil and plants, as in chemistry generally, is important and can be quite mysterious. Basalt has unusual characteristics, including a property called paramagnetism, which is having a set of physical characteristics including: (i) a slight reaction to a magnetic field, and (ii) no permanent magnetism or independent generation of a magnetic field. It is fascinating and can be very mysterious because magnetism itself is not well understood. Paramagnetism is a form of magnetism that is difficult to explain quickly, other than to say that it is a weak and intermittent form of magnetic effects from external magnetic forces.

Broad-spectrum rock dust with high paramagnetic qualities builds soil's ability to sequester carbon long-term. It is one treatment that is not in the conventional soil science library, as it is not a proprietary product that can be sold for a high profit. It does not fit the chemical soil testing mindset. As mentioned, soil testing is an area that needs to be more universal and practiced with regularity for the greatest impact under The Get Real Program.

Adding basalt dust to the soil changes the paramagnetic capacity of soil. In even extremely depleted soils, somehow basalt has positive effects on plants. Curiously, oxygen is highly paramagnetic, as are many other materials such as some volcanic rocks. One researcher specializing in low-energy energy fields, internationally renowned entomologist and ornithologist Philip Callahan, says high paramagnetism creates better soil fertility and productivity. Tests have showed increased plant growth when basalt rock dust was put in a plastic bag under the plant, and it is theorized that some of the benefit may be due to the paramagnetism of the basalt. Nobody understands all the details of how plants interact with the electric and magnetic fields of the Earth and why certain plant chemistry works better with paramagnetic basalt additives. Unfortunately, it's an area that doesn't get much research.

The Get Real Program believes in the value of paramagnetism and its positive effects on photosynthesis, which justify increased investment in research to learn all that can be gained from this incredible and mysterious phenomenon. The Program feels that only the surface has been touched from the depth of this area of science. As we continue to evolve the richness in our soils, it will be exciting to see the increased impacts from paramagnetic materials.

TECHNOLOGY TO THE RESCUE

Rock dust is a key basis of good soil, and the technology to create it and to spread it on land is known and can be improved. We must think boldly, changing the way our species does things, in order to remineralize the world. Renewable energy is great for grinding and transporting rock dust around the world. Solar photovoltaic and wind energy are variable and hard to rely on for constantly available electricity, but they are great for grinding rocks.

Imagine a wind-powered rock grinder grinding up basalt until it is like fine flour, covering large areas of land with rock dust so fine it sinks into the soil with rainfall and travels to wetlands and the oceans, revitalizing them too. Imagine a fleet of ocean-going vessels transporting basalt rock from coastal quarries to regions lacking their own deposits – ships powered by wind, wave, and solar energy to grind large rock into smaller particles enroute. With rock dust particles so fine, winds could pick up a portion of it, but it could also be blown onto the surface of the ocean, remineralizing the ocean as the ship transports the bulk of the dust, possibly processing some of it into pellets along the way for easy handling at its final destination.

At 100 tons per hour, operating nonstop, it would take six weeks for a wind-powered vessel containing 100,000 tons of rock to grind it all. Solar photovoltaic power would take 24 weeks to achieve the same results, since solar power is generally less efficient than wind power by a factor of about four times due to only being capable for a few hours a day. Such vessels would look nothing like conventional ships and would be so large as to never come into harbor. They could stay offshore while smaller barges would bring the rock pellets to the land, or possibly even use airships that could access areas with no roads – which is most of the world. Visit www.fullofideas.com for more information on this and other innovative ideas. A *vessel* could really be a cluster of floating structures supporting a vast array of solar photovoltaic panels to grind rock dust, along with wind and wave energy. (See the later chapter on smart energy options or go to www.fullofideas.com for more details.)

A NEW ENVIRONMENTAL APPROACH

Modern nations already spend vast amounts of their revenue on military defense. Why not move to spend a similar amount on saving the planet from ecological devastation?

Remineralizing the earth would be humanity's greatest project – one that is doable and aligns perfectly with carbon sequestration efforts. Unlike chemical fertilizers, there is no upper limit to basalt rock dust application, as the nutrients are not water-soluble and will not overwhelm the soil chemistry. It takes microbial life to release the nutrients.

Economically, it is desirable to apply small quantities of rock dust – less than one tenth of a percent of topsoil mass or about one ton of basalt per acre. More can and should be applied over time. The finer the rock particles, the more surface area per pound and the fewer materials needed to achieve the desired short-term remineralization. However, as soil life will deplete nutrients over time, additional rock dust will be needed, eventually. Flour-like rock dust is great for soil nutrients but not so good for soil structure and air and water infiltration, thus small quantities are important. Basalt rock dust has good weathering properties that help sequester carbon in a chemically stable form. The limited research that exists shows positive results.

Fortunately, the basalt mining industry currently uses basalt to make coarser products, separating and stockpiling the finer rock dust. This means there is a large supply of fine rock dust ready to kick-start remineralizing without delay. This stockpile could remineralize around a million acres in the U.S., but new mining would be needed to truly make a difference in the world's soils. To those reading this book, your support for the nonprofit organization Remineralize the Earth (www.remineralize.org) will help build momentum for larger scale use of basalt rock dust.

Climate activists should support rock dust crushing and delivery systems, powered by renewable energy. It may seem strange, but using wind-powered and solar-powered rock carrier ships may be part of the best possible carbon sequestration solution. Much of the world has no roads, and new design airships may offer an innovative means of distributing rock dust to restore fertility to forests and all other areas, no matter how remote. For more information on airships and other updates and links, visit the founder's website: www.fullofideas.com. To see one company's airship prototype, visit www.lockheedmartin.com/en-us/products/hybrid-airship.html.

The Get Real Program is a way to reverse the dangerous demineralization and soil depletion that presents a far greater threat to humanity than atmospheric CO_2, and fixing the soil will also solve the CO_2 problem in a positive way! Onshore, expansion of the rail network and new methods of transport (to be discussed in the smart energy options chapter) are needed to mine, move, and distribute the enormous quantities of remineralization feedstock. The world is a big place and doing the work of an ice age is a daunting task. In truth, rail is just one of many things that we must reimagine and develop. If we can move billions of tons of rock dust cleanly and efficiently, we can change the trajectory of our soil from depletion to rebuilding.

This book is not a half-hearted proposal but a bold manifesto for a dramatic turn for the better. It focuses on a big solution to climate change—one recognizing the need for more sustainable, renewable energy—but it does not get caught in the proverbial weeds of trying to eliminate fossil fuels.

ROCKS MATTER!

Unfortunately, some scientists do not give soil much credit for its true ability to sequester massive amounts of carbon if treated and managed properly. Some climate activists tend to overlook two of the most important things on Earth – soils and the ocean. Both are vital parts of the carbon cycle and both are currently demineralized.

Scientists generally cling to the mantra that only drastic cutbacks in man-made emissions can slow the rise of atmospheric CO2. They should look at how best to support natural systems that can actually lower CO2 dramatically while we still use oil and gas instead of impractical plans to just slow the rise with draconian cutbacks. They should be focused on remineralizing!

Most rock dust proponents are often too timid to promote even the use of small quantities of fine basalt rock dust left over after quarrying, much less the massive mining efforts that are actually needed. Their estimates of potential carbon sequestration are a fraction of the potential with a truly effective, global remineralization campaign. The world needs rock dust. Lots of good research shows it to be effective, although some loud voices dismiss it in favor of protecting profitable chemical products and/or out of ignorance of the importance of rock minerals. Sadly, many people simply do not know of or see the earth's potential to lower atmospheric CO_2 levels with improved soil carbon sequestration aided by remineralization!

CARBON SEQUESTRATION SOLUTION: ROCK DUST

Transforming soil into a carbon sponge requires restoring the soil to health, but this takes money. Fortunately, using carbon sequestration funding provided to charities to restore rock nutrients to the soil will produce a long-term climate solution and help pay for more vigorous plant growth. Remineralize the Earth (www.remineralize.org), a non-profit organization based in Northampton, Massachusetts, is a great source of information on this topic. Their book *Geotherapy: Innovative Methods of Soil Fertility Restoration, Carbon Sequestration, and Reversing CO_2 Increase* is scientifically written and heavily footnoted. It thoroughly explores the remineralization of soils and forests as a key strategy to stabilize the climate. Please visit their website, become a supporter, and sign up for the newsletter to help promote this vital solution to the climate crisis and the greater soil crisis.[4]

A chapter in *Geotherapy* cites a Panama rock dust study that showed an almost eightfold increase in biomass for acacia trees is possible in only five years, which suggests the possibility of storing almost as much as eight times more carbon. The chapter notes that "Growth rate differences expressed in terms of biomass increase would be much higher than the 2.17-fold increase in height (i.e., plant height) found between local soil (alone) and fresh basalt (i.e., soil with basalt dust added), due to the thicker trunks and stems, more abundant leaves, improved nutritional status and hence leaf weight per unit area in soils supplemented with fresh basalt rock powder." The advantage of basalt dust addition is a clear potential advantage and is part of The Program.

Basalt powder can restore soil fertility and greatly accelerate tree growth on Panama's otherwise impoverished tropical soils, and it has the potential to provide abundant returns when applied

---

[4] Edited by Goreau, Thomas J., Larson, Ronal W., Campe, Joanna, ***Geotherapy: Innovative Methods of Soil Fertility Restoration, Carbon Sequestration, and Reversing CO_2 Increase***, Boca Raton, FL, CRC Press, Taylor & French Group, 2015.

elsewhere as well. Long-term experiments in Europe, released in the 1986 showed that in a forest where pine seedlings were remineralized, after 24 years the wood volume was four times higher than in the untreated area. One application lasted 60 years!

THE BIG WIN OF BASALT AND ROCK NUTRITION FOR SOILS AND THE FUTURE

America has a role in preserving this Earth, but the need is far greater than what America can impact alone. The big win for life preservation for the distant future would be a global program to remineralize every arable piece of land in the world… all the forests, grasslands, and deserts.

There's an effort right now called the "Great Green Wall" in Africa that is trying to stop expansion of the Sahara Desert by planting trees that can survive in arid environments in front of areas of advancing desert conditions which are threatening Northern Africa. It would be great to have basalt rock dust added to the seedlings to boost their growth and survival in the effort.

We must go global with our influence to create real change for our environment and enhance the health of Americans and everyone through the food we eat and the air we breathe. We must be the example to the rest of the world that something can be done successfully to manage carbon emissions, while also saving our soil and ultimately, life on Earth. The time to invest and take action is now. Let's create change together for generations to come. Join The Get Real Program community at www.getrealalliance.org to find more resources and information on saving our Earth.

CHAPTER 4

MAGNIFICENT WONDERS IN BIOCHAR

While rising CO_2 levels alarm many, the depletion of the Earth's resources is perhaps far more daunting and dangerous. Yet, this is largely unspoken by the activists who claim to "care" about the sustainability of our environment. It is the farmers who can recognize the real threats and are ringing the alarm bells, because they know what is required. Quite frankly, investment, innovation, and change of man's practices, focused on real solutions for our soil and biosphere, will have much greater direct impact on this Earth than reducing gas emissions from cattle and limiting oil and gas combustion.

As sprinkled throughout this book, The Get Real Program focuses on the real issues of our environment by offering biochar as part of a comprehensive solution towards environmental prosperity. Converting biomass that would otherwise decay and oxidize back into the air into biochar is a priority to control CO_2 levels. Creating a pathway for wide distribution of biochar is also key to our success in revitalizing our soils and supports other gains in less politicized aspects of our planet's vitality.

Man is taking more nutrients from the land than are returned. Through man's current practices, such as mining, major nutrients of nitrogen, phosphate, and potassium are used wastefully with over half going into the sea or air. Mining is like living on money from a savings account. It is easy as long as there is money in the bank (or minerals in the earth). No matter how big the bank account, drawing out more than the interest earned without making any deposits will eventually deplete it. Sustainability involves living on less than one's income, which is harder to do than living on stored wealth or resources.

It is easy to live temporarily on your savings as long as they last, but eventually those times come to an end. Man has been living on stored nutrients in the soil and getting away with very destructive soil practices that have washed away billions of tons of topsoil. We have reacted to depleted soils by breeding plants that use fewer nutrients and that produce bulk instead of quality. Plants can produce lots of carbohydrates using nothing but air, water, and a tiny amount of rock nutrients. But those plants do not provide enough nutrients for people or animals to thrive on, only "empty carbs."

Living sustainably with the soil doesn't mean that it can't be more productive. It only means that it may cost a little more to get the soil revitalized so that photosynthesis can work its miracle of making matter from atmospheric elements and sunlight. Making billions of tons of biochar can boost soil life and prevent the loss of vital nutrients through leaching and nitrogen losses. Oh, and there is more...

WHY BIOCHAR?

Working with Biochar will lead to a net reduction in CO_2 emissions from natural sources with more than half the total carbon contained being locked up in underground/soil-enhancing biochar for centuries instead of rotting and oxidizing within a few years. Making biochar can sequester carbon for centuries, but also can create an environment in the soil where soil life can flourish and sequester far more carbon in a stable, long-lasting form. Most importantly, ongoing biochar distribution can permanently remove carbon from the carbon cycle! Broad distribution can also enhance the air we breathe, improve the food we eat, clean up the water, and so much more.

Many activists see carbon as something that must be disposed of instead of something that plays a role in a healthy ecosystem. However, farmers know these schemes to pump CO_2 deep underground are wasteful. The carbon cannot do anything positive underground, except in the case of depleted oil fields where it can push more oil out (and this is a good form of sequestration as well). The Get Real Program uses biochar as part of an overall solution that will work within the natural carbon cycle to create real change and impact on our environment and our lives.

Biochar is very porous and thus has a large surface area. A consequential aspect of biochar is its capacity to absorb large quantities of water. This benefit can result in great habitats for soil life. By itself, biochar has no nutritional value for plants as the carbon is biologically stable. But when primed with plant nutrients through compost, manure, or fertilizer, biochar becomes a stable source of nutrients for soil life and plants. Elements are attracted to carbon and bind to it, making them unlikely to leach away.

Also, Biochar's attractiveness to elements and its porous surface area make it an excellent filter to remove harmful chemicals from water. When making biochar out of crop residue in a processing plant or facility and then efficiently distributing it to be incorporated into soils and newly reclaimed arid lands, carbon is removed from the carbon cycle permanently and put into the soil where it is beneficial.

Nothing is worse for the environment than bare, sterile land that is a high net carbon emitter. As an example, it would be far better for improved and more extensive soil life to consume soil carbon aggressively when the corn crop isn't growing (between seasons) and in the middle of the rows (during the season) with little change in current farming practices, as is proposed by The Get Real Program.

In some cases, biochar from other areas will need to be brought in to reclaim regions of desert and regions that have declined and host only sparse vegetation. To succeed in turning the world's farms and ranches into active carbon sequestration sites while lowering CO_2 levels, we need to take reasonable actions. The changes being discussed will have added positives beyond carbon. They will also increase the nutrition of food and increase profitability of the farming properties beyond the investment we suggest from carbon sequestration charitable funding.

THE HISTORY OF BIOCHAR

Looking at history, it seems that ancient peoples in Latin America learned how to make and use biochar. People of the Amazon basin used biochar to improve land from a relatively low fertility clay-based soil to a perpetually fertile soil mixture known as "terra preta." There are many books and videos about the amazing soil. The book *Terra Preta* by Ute Scheub, Haiko Pieplow, Hans-Peter Schmidt, and Kathleen Draper, among others, is a good resource to learn more about biochar.[5] The comprehensive book *Geotherapy* has detailed information about biochar and rock dust. The following selection from *Geotherapy* is the concluding passage of a paper on biochar published by CRC Press:

> "In conclusion... Biochar as an industry could become as large as the fossil fuel industry. Biochar ... could approach that of the world's largest industry, the food industry, with which it can be in more than full harmony. Biochar offers hope to reverse global warming. The hope biochar offers is needed in society today."

[5] Scheub, Ute, Pieplow, Haiko, Schmidt, Hans-Peter and Draper, Kathleen, ***Terra Preta, How the World's Most Fertile Soil Can Help Reverse Climate Change and Reduce World Hunger***, Vancouver BC, Canada, Greystone Books, 2016.

THE MAKING OF BIOCHAR

Biochar is biomass that has been burned at a low temperature, usually in an oxygen-deficient fire, so that only pure carbon remains. Pure carbon generally doesn't burn or oxidize at less than 1000 degrees, so it is stable for thousands of years. Making biochar from plants is a process similar to making wood charcoal. By burning the volatile carbohydrates and evaporating the water from plants, large amounts of pore space are created in the remaining fibrous structural material. An industry making biochar out of crop and land-clearing/weeding residue removes carbon from the carbon cycle permanently, after efficiently putting it back in the soil where it is beneficial.

It is counterproductive to burn long-lifespan plants, such as trees, to make biochar. A living tree sequesters carbon without releasing it back to the environment on a yearly basis and may live up to a hundred years or more. Ideal candidates for biochar are undesirable plants, like weeds, that take up space and nutrients where good grass could grow, and deadwood, as well as trees harvested to create healthy forests, not overgrown and creating huge wildfire risks.

Today, production of biochar is low despite there being a number of biochar-making systems and an increasing awareness of biochar's benefits. All Power Labs is a small business that sells and develops systems to burn biomass for power and process heat while producing biochar. It is worth visiting their website to learn more about the potential of biochar applications. The following is an excerpt from their website:

> "When fueled by sustainably sourced biomass, power generation with a Power Pallet is a carbon neutral process. However, the Power Pallet can also stabilize a portion of the biomass carbon, which would have been released in natural decomposition, into a carbon-rich waste byproduct –biochar. When

> mixed with soil, the carbon in the biochar can be sequestered from the atmosphere for centuries or more, making the Power Pallet … carbon negative."[6]

Almost all biochar is currently made from trees, although the International Biochar Initiative has research papers on using other waste streams as feedstock. You can learn more about this initiative at http://biochar-international.org/. There are hundreds of millions of acres of unmanaged or under-managed forest in the United States that contain mostly dead trees. One great feedstock source for biochar supply is deadwood in government-owned forests, which will require government to take action. Grasses and weeds can also be great feedstocks for biochar as they are widely available and quickly oxidize back into the atmosphere as they die.

Deadwood stores carbon, but the carbon is headed back into the air in a much shorter amount of time when compared to biochar, which locks up carbon for a century or more. If there is a forest fire, carbon from deadwood is quickly returned to the air, and living trees will burn due to the added heat. Many forest fires would not spread or do so much damage to living trees if it were not for all the dead trees and debris that are in most of our forests in the U.S. today.

One of the less desirable things about making biochar is that most processes involve burning the feedstock material in an oxygen-limited environment. This is done to control temperatures and burn off only the volatile hydrocarbons and impurities, leaving behind only pure carbon structural elements. Ideally, when partial combustion is begun, the released gases are used for fuel, adding value to the process. However, burning these gases releases a lot of CO_2 into the air that minimizes the carbon sequestration.

When making deadwood into biochar, the produced combustible gas can be run through a power plant to make power. This is a great

6 https://www.allpowerlabs.com/carbon

improvement, but only works for dry material. Other processes which efficiently convert wet material and capture resulting gases for other uses can make the production cylcle increasingly carbon-negative.

The Renewable Energy Connection

Now the question is whether to biodigest the biomass for biogas before charring the digested material. The answer may depend on the situation. Providing reliable energy to the economy from renewable resources should be a priority. To do this, many biodigesters (more on this topic is in the chapter on smart energy options) are needed to convert and utilize the energy of photosynthesis as efficiently as possible.

The investment will pay off in multiple ways. For instance, gathering up corn stalks and other crop residue is beneficial on two fronts. First, the gathered materials can be made into biogas that will provide renewable energy. Second, biochar can be produced out of the remaining biomass. There is on average five tons per acre of wasted dry matter. Multiply this figure by 100 million acres and that's 500 million tons of carbon that could be removed from the carbon cycle that would otherwise be headed into the atmosphere. Converting the digested corn stalks into biochar, as discussed in greater detail later, increases the conversion rate and can allow wet feedstock to be used. The production of 300 million tons of biochar would make up a significant portion of America's fossil fuel burning CO_2 emissions.

Additives of Biochar

Biochar can be mixed with other nutrients including mined fertilizers. There is a need to restore soil and plant life to arid areas that are headed towards being total deserts. These at-risk areas are usually nearly devoid of soil carbon-containing components and have little soil life. Because of the lack of organic matter, the soil is often nearly sealed against water infiltration and storage, preventing most plants from living.

There is an effective practice in Africa of making small holes in hard desert soil and putting a little organic matter and a seed in the hole. The hole catches the rain when it falls and creates a micro-environment for a plant to grow. The process could be improved with the use of nutrient-laden biochar and rock dust. By adding these materials to small water catchments, photosynthesis and carbon sequestration can be stimulated in lands that are currently carbon emitters!

REGENERATING WITH BIOCHAR

The Get Real Program advocates for large-scale production of biochar and its distribution and application to the land to regenerate rich soils. The result of these actions is the long-term removal of carbon from the carbon cycle while enriching the land and preventing the leaching of vital soil nutrients. Biochar that has been loaded with nutrients is a great environment for a wide variety of soil life.

Soil with lots of biochar along waterways and drainage areas can soak up nutrients that are currently leaching off the land. It also filters the water and absorbs fertilizer and other things out of the water to make it cleaner. This keeps them from going into the waterways and areas where water runs off; thus, it can keep toxic algae from being created. At current rates of use, there are deposits of phosphate and potassium to last many decades. This sounds very good, but in aiming for sustainability, we should have centuries in mind rather than decades.

The concept of Net Present Value, which is widely used in business, gives reserves a century or two in the future no present value due to the effect of interest. Because of this, a mine owner will mine as much as possible as fast as possible. Fertilizer companies have encouraged high rates of use of their products, even to the point of seeing much of it go, unused, into the ocean. Once these mineral deposits run out, the miner is out of business and so is society. Farming production rates could plummet if other mineral sources are not developed.

Producers may be unable to support a population that was allowed to grow to an unsustainable size.

BIOCHAR AND WATER

Putting millions of tons of biochar into arid soils will increase their water absorption and storage ability. This all sounds great, but it may reduce the amount of runoff water that goes into lakes like Lake Mead, which provides water to so many. As we know, having more vegetation growing in the area increases plant transpiration, which may trigger more rain. Geotherapy on a large scale can change things in unforeseen ways. In general, arid land doesn't get much rain due to low plant transpiration or water evaporation from leaves. Rainforests actually create their own rain by increasing the humidity in the air above the saturation point where rain falls. Grassland has a lower transpiration effect, but it is still significant. Bare ground that soaks up very little water evaporates very little except right after it rains. The rain that does fall carries away more soil into rivers, lakes, and then, the ocean.

It is ironic that converting near-desert land into grassland may not reduce the Earth's average temperature. This is because a desert has such low humidity that it doesn't trap heat. After the daytime solar heating, the heat radiates back into space at night. The nighttime temperatures drop very low because there is no water vapor holding in the daytime heat. This results in an average temperature that is quite different from either extreme.

THE BIOCHAR INVESTMENT

The potential of biochar is fantastic and limited only by funding. Unfortunately, many farmers are currently in a state of economic depression, unable to make long-term improvements that require up-front cash. Sequestering carbon out of the air and keeping dead biomass from oxidizing back into the air are huge tasks. But they are necessary in the medium-term and in the long run if we want to lower CO_2 levels while still using oil and gas.

Biochar is now used on a small scale as current methods of making it are expensive. Carbon sequestration funding could support a large effort by contracting the younger generations and other manual labor to regenerate the desolate regions of the western United States. Such an effort could avert disaster in a region that is becoming a desert due to bad policies and neglect. It could also provide fulfilling work for the many non-violent criminal offenders.

There are so many tons of deadwood per acre in national forests that most of the produced biochar will need to be taken to other areas, with only a small amount being mixed into the forest floor. Since most of the western timber is owned by the federal government, programs can be set up that benefit the long-term interests of the nation and our forests. Investing in this new way to harvest and improve the forests will bring big benefits to the nation in lowering carbon emissions from decay with nearly-permanent removal of carbon from the carbon cycle. It will also make the forests healthy, resulting in faster, more sustainable growth and sequestration of more carbon.

The government needs to take greater action and fund forest restoration of federal lands. We have seen so much waste in government spending, yet key areas requiring maintenance, like federal lands, are not receiving proper care and investment. If the government owns assets, it cannot just ignore them to the detriment of taxpayers. Something has to change, and soon!

A carbon sequestration payment system will funnel funds into the countryside, creating a need for more people to live and work in rural areas. Biochar production and distribution can be a major industry along with making biofuels. An era of living sustainably and on largely renewable energy puts the cities at a disadvantage.

If we are going to recycle our food and other waste back onto the land, it is better to be closer to productive agricultural land.

Because utilizing biochar results in near-permanent removal of carbon from the carbon cycle, farmers applying biochar deserve to be paid more money to produce it than what is paid for their crop. Farmers' financial situations don't allow them to obtain biochar-making equipment, so charitable funding should be used to buy the equipment for centralized processing, as over time, many tons of carbon will be sequestered in a beneficial way.

Things like carbon sequestration and biochar production offer new revenue streams that work better with larger-sized farming operations or co-ops. We need to rethink our agricultural practices so as to be more sustainable and provide a good living for rural people. It is ridiculous that, in most cases, a farmer has to have outside income to be able to maintain a farm. The average farm is a net carbon emitter as it uses so much fossil fuel and doesn't treat the soil well to sequester carbon. We can invest in the farmer via co-ops and shares in centralized facilities (just like farmers have invested in mills, gins, and grain elevators in the past) for sustainable positive change.

NECESSITY OF INNOVATION

It will take a massive innovative effort to fundamentally change the soil for the better.

Let's Get into the Weeds... Biochar Solution

FOI Group, LLC, has developed a new innovative method of producing biochar in a processing facility using residuals of green grass and weeds. Weeds are a major problem in agriculture. Many billions of dollars are spent on harmful chemicals that are applied to fields around the world in hopes of controlling weeds. Over the years, weeds have developed immunity to many chemical herbicides, making chemical control of weeds in crops costly and ineffective.

Man often underestimates the ability of nature to adapt to threats. Organic and regenerative farmers and researchers have lots of evidence of the harm to soil life and sometimes to people that results from the use of the poisons in our current herbicides. Weeds resistant to the most effective broad-spectrum herbicide, *Roundup*, are already spreading across the country and *Roundup* is being accused of being a carcinogen – very harmful to wildlife and people. The industry's reaction has been to genetically engineer crops to be resistant to other more harmful chemicals that can control these weeds, knowing full well that in time the weeds will develop resistance to them.

Modern conventional agriculture is entirely dependent on chemicals to prevent weeds from taking over fields and rendering conventional farming impossible. With drastic action, weeds could become an asset instead of a liability. Weeds perform a valuable function of using photosynthesis to feed soil life so that it can sequester carbon in the soil. You can go to www.fullofideas.com for more information.

Soil microbial life will deplete soil carbon reserves for food, resulting in the emission of CO_2 into the air. To make the soil a net absorber of CO_2, major changes to agricultural practices are necessary. Action to sharply cut soil erosion is needed in addition to action to cause the soil to soak up more rain, preventing flooding and leaching of soil nutrients into the waterways. Biochar can meet some of these needs by improving water infiltration and storage as well as retaining soil nutrients, which will also allow for bigger, healthier primary crops, cover crops, and new organic (weed) activity, which can sequester more carbon.

Digesting Crop Residue

Left-overs and whole plants are beautiful when it comes to gathering crop residue (i.e. harvested corn stalks) to be converted into biochar. Transporting crop residue along with harvested crops to a local biodigester system where energy is extracted from it before it is dried and oxidized by solar energy and atmospheric oxygen is an efficient way of sequestering CO_2 from waste. The biochar is then distributed back to the farms (as a back-haul item which is efficient) and transported back to the soils by the gathering system to be incorporated directly into the soil. Adding only the residual weight and a backhaul commodity just slightly increases the carbon footprint, cost, and complexity.

Adding livestock to the system helps make better use of crop residue, especially if cover or companion crops boost forage production. One of the problems with mega farms is that many things, like keeping cattle, are hard to do when operations are spread out. Tasks relying on mega-sized equipment to cover hundreds of acres a day with just a few people make things difficult.

Driving in North Texas recently, the Founder observed a great deal of bare dirt, often with gullies washed in it, during the winter. Crop residue has oxidized away leaving the soil to demineralize and lose organic matter. One cause of this is a farmer planting a single crop of GMO corn in fields and leaving the middles of the rows bare. That means that most of the soil has nothing growing on it in summer and so it is a net carbon emitter all year. An efficient biochar distribution system can be part of the solution here.

Let's Talk Renewables… and Biochar

The Get Real Program has many ideas that advance thinking on biochar. For instance, the use of focused energy to char plant and weed material at a processing facility near crop collection points (such as alongside grain elevators or rail storage depots) can change the way biochar is made for the better. Such a facility could operate autonomously during times of adequate wind or solar energy availability, slowly making biochar that can be collected an distributed back to farms put into the soil.

All soils can use biochar to improve water and nutrient retention and improve water infiltration, but many areas don't have available wood to make biochar using the current kiln process of partial combustion. For instance, in a harvested corn field there is so much biomass to convert. In a sparse pasture or non-irrigated tract, however, collection can be fast. Harvesting/collecting the biomass can be automatic in many cases, but for desert reclamation, it may be better to have people selectively cut or trim the plants and send them to a facility. Some areas such as deserts, where there is little feedstock, will need to have biochar imported.

In the desert, the program would involve digging water-catching holes and placing biochar and rock dust in the holes to provide a habitat where desirable plants could grow to help revegetate the desert.

The use of solar biochar facilities to process inedible and leftover forage after cattle (that are being holistically managed through constant movement to new pastures) move on offers a way to turn undesirable vegetation into biochar. This is most feasible when the pasture has a lot of undesirable plants not suitable for grazing, and it is desired to increase the grasses. Such fields can't support many days of grazing, so the cattle don't leave a lot of manure. It may be that a pasture has mostly inedible plants, making it a good source for biochar feedstock.

Heating, drying, and charing wet biomass allows sea plants and marine wetland plants to be used for biochar, preferably after they have been biodigested for biogas. The world has a large appetite for natural gas, and it will take a massive effort to build enough biodigesters to augment the supply significantly. Investment in biochar facilities would fund the removal of carbon from the carbon cycle, improving water absorption and retention on land, which increases plant and tree growth without increasing the demand on local water supplies. One of the big problems is soil that has very low water infiltration and holding capacity, where rains just run off it, usually carrying soil with it. We need to add massive amounts of biochar to the land to make it a sponge for rain everywhere.

About or over $200 per ton could be paid to operators of a facility. Many farmers and ranchers would want to own shares in biochar facilities. Biochar has long-lasting value for fertility and nearly permanently removes carbon from the carbon cycle. So, it is worth more than carbon compounds in the soil as far as a permanent sequestering.

The Get Real Program proposes building numerous facilities a year until initial demand is met, but the rate and total amount would depend on the goals and future methods that may be developed to make biochar. It will be necessary to have fixed-location solar biochar facilities located next to large biogas digesters to process the solids in the effluent. The effluent from a biodigester consists of rich nutrients as well as indigestible fiber that the anaerobic bacteria could not break down. The fiber is of low value to the soil and would be more useful if made into biochar.

These systems may also be used in conjunction with algae farming operations to provide water and nutrients to the algae that would also be fed with CO_2 from the digester. Algae is a great source of biomaterials and biofuels and needs to be a big part of a sustainable and renewable America. The new integrated farming operation could have multiple activities that go on year-round, providing for society in many ways. Many of today's operations grow one or two crops a year, causing soil loss and nutrient mining and using lots of chemical inputs. They then sell at wholesale prices. The regenerative farm will produce a variety of food and the farmer will get paid to sequester carbon.

Full of Ideas is working on a patent filed on a biochar maker. See www.fullofideas.com for more information on biochar methods.

Solar and Biochar Processing

In arid lands that are tending towards becoming desert, it is desirable to create micro-environments for grass to flourish where biochar and rock dust are placed in water-catching indentations to prevent rain from rushing off down gullies. Ultimately, the grass needs to be holistically grazed to increase its vitality, but that may be impractical in arid areas, especially at first. Ideally, grass needs to be cut back after frost to allow sunlight to reach the new growth in the spring. Efficient biochar processing and a distribution method to incorporate it into the soil could effectively create new grass structures or add biochar to existing pods.

Such arid areas will be low in pounds of biomass per acre, but it is important to get more vegetation growing on this land with an ultimate goal of supporting managed holistic grazing. Most of this land is owned and controlled by the federal government, and there are many different opinions as to how it should be best used and maintained. There is a lobbying group that promotes initiating NO change to the arid ecosystem and is happy to see the land degrade into desert or near-desert due to natural degradation of un-grazed land.

In countries that burn wood and other biomass for fuel, biochar-producing wood stoves need to be implemented as part of a carbon sequestration scheme. Providing devices for cooking and making biochar from waste would also be good. All countries should be supportive of this method since CO_2 is a global issue and many disadvantaged areas desperately need biochar to hold more nutrients and water in the soil for better crops.

The Innovative "How" of Biochar Distribution

Given that most of the world doesn't have roads, and in many cases can't economically build them, a new generation of modernized and even autonomous airships to do the vital work of spreading rock dust and biochar seems like the best solution. This idea may seem far-fetched, but the objective of The Get Real Program is to promote new ideas and areas of study on how to make the world a better place for all.

The biochar could be transported by air to areas that need to be regenerated and are becoming desert but have no navigable roads.

The other part of this solution would be renewable-energy-powered lighter-than-air transport craft to take logs and biochar out of forests and bring rock dust into forests and other remote areas. Harvested, dry wood chips could be used in a biochar-producing energy system to produce power to propel the airships long distances with renewable biogas or other fuel being used as needed.

Moving things by airship can be very fuel efficient by partially powering the vehicle with a wood energy system that heats the helium for more lift, giving the craft more carrying capacity for a given amount of helium. Solving the problem of removing dead trees from dying forests will take an innovative approach because so much of the areas of interest are far from roads. It is wonderful that America has these roadless wilderness areas, but it's unfortunate that they are so unhealthy and mismanaged, resulting in the risk of mega-fires.

It is often thought that areas far away from man exist in an ideal state. But nature is degenerative; the Earth's minerals are leached away over time until restored by volcanic or windblown rock dust. Man intervening in the forests can make a big difference for the better. Remineralizing the forest with rock dust and removing excessive deadwood can improve America's forests.

Though the idea of using environmentally friendly airships is appealing, there is a problem with widespread adoption and funding needs. Shortsightedness, an unfortunate human trait helping to cause many of the problems this book addresses, has resulted in a helium shortage. A stockpile of helium started by the federal government in 1925 is being rapidly depleted. Many industries (i.e., medical, semiconductor manufacturing, etc.) are concerned about helium shortages and rising prices. Analysis of the global helium supply is complicated. Helium is found in geologic natural gas reserves, and there are recent efforts to extract more helium from these reserves. It is possible that the helium supply will be adequate once new extraction sites start up. Hopefully, helium availability will increase so that a fleet of airships can be deployed to save the vast western forests of the United States and to remineralize many other roadless areas.

The use of hydrogen for lifting non-human loads with airships should be considered. With modern flight technology, it would be possible to remotely pilot many of the crafts doing the tedious work of moving billions of tons of biochar and rock dust. Such developments would allow for the construction of a massive fleet of aircraft to remineralize and revitalize the entire world.

One solution could be tethered lighter-than-air crafts that can hover above the forest and lift dead trees up into the craft to be chipped and made into biochar. A craft could be large enough to house workers and provide hoists to raise and lower them to the forest floor to cut dead trees and remineralize the soil around living trees with rock dust. The Get Real Program is practical and realism-oriented in the solutions proposed, but as engineers, we believe it is important to have an inventive and forward-looking mindset for innovation. See www.fullofideas.com for more on airship solutions.

Aerial Harvesting

Traditional logging is destructive and dangerous. An aerial harvesting method can offer big improvements. Grabbing a tree with a grapple near the top allows the trees to be lifted out of the forest without damaging surrounding trees as happens with cutting. Lifting the tree up to a work deck below the airship creates a safe workplace to trim off the limbs and turn them into wood chips to make biochar. Lightweight bandsaw lumber mills can cut off unusable wood for onsite chipping while saving the higher value wood to be cut for lumber. There will be many trees that are too far gone to be anything but biochar, but quality timber should be used for more valuable purposes.

Angled wind turbines could provide lift and power to the craft. Using solar energy to partially power the aircraft, while heating the helium to increase lift, is possible. Deadwood contains lots of energy that can be extracted as wood gas during the charring process. Wood gas can power the chipping process and leave a lot of the chips to be used as fuel for transport airships. The ship can gradually turn the wood chips into biochar as it travels from the dead tree harvesting site to an application location. The key is to innovate our thinking for creative solutions.

EMPTY ARGUMENTS

Biochar has many detractors. Many activists want to solve the CO_2 crisis with solutions that involve extreme sacrifice and deprivation by man. Their claims against biochar, a solution to the CO_2 crisis that does not involve such sacrifice and deprivation, are based on deeply held convictions and misconceptions rather than science. Some claim that biochar biodegrades and doesn't last for centuries, yet true biochar has to be heated to over a thousand degrees to burn or oxidize.

Under normal conditions, biochar is stable for the foreseeable future. Naysayers give the soil little credit for sequestering carbon, so the fact that biochar increases soil life activity gets very little credit. For many people it is becoming practically a religious belief, based on shallow mass-media analysis and repetition, that the only solution to the climate crisis is to drastically and rapidly phase out fossil fuels regardless of the cost and loss of quality of life.

The Get Real Program argues for using fossil fuels to fund a carbon-negative future for the world that will start soon rather than simply slowing the rise of CO_2 over a matter of decades. There are lots of facts to back up The Get Real Program, but the "religious fervor" of opponents of this proposal and its parts makes it impossible for them to listen to logic or accept proof. An additional factor is the human tendency to look at the time of our youth as being better than today. The reality is that things are getting better over time (i.e., increasing fuel efficiencies and renewable energy technologies), though now there are, of course, things that need to be improved. One is the treatment of the soil, which requires continued improvement to fully adopt the improved methods used by a few. Biochar usage programs are rapidly gaining credibility; it is almost ready for widespread adoption. Let's help continue to spread the message and invest our resources towards real solutions proposed by The Get Real Program and then…

LET'S GLOBALIZE

The new renewable-energy-powered airships would have the potential to help in countries where the lack of established roads is a limiting factor. The destructive burning of grassland and forest in many areas of the world needs to stop. American private funding of charitable organizations to support the sequestering of carbon will be very effective at sequestering more carbon than America emits; however, a worldwide effort is needed to really change things for the better on a global scale. A global effort to make lots of biochar can transform both the climate and the soil for the better. America has no choice but to lead on the issue of our environment. Let's start today!

BIOCHAR IS THE ANSWER

We are entering a time of crisis. Real change must be made to ensure the healthy survival of plants, animals, and people. The good news is that change is doable and will generate long-term gains for society. Paying landowners to sequester carbon will revitalize rural areas that are now in depression due to low prices and the rising costs of conventional agriculture.

We can bring our farming to the next level by building up the soil with biochar. This will minimize the waste of fertilizer due to runoff. Biochar can capture significant amounts of the fertilizer that is wasted when leaving the property. Biochar can permanently remove carbon from the atmosphere and carbon cycle, while also improving water retention and extending our fertilizer reserves. This will ensure America has a sustainable food supply while restoring the deserts and making tree planting more viable.

It is amazing to consider how making and distributing massive amounts of biochar in the Mississippi watershed could sharply lower the amount of phosphate, nitrogen, and potassium traveling to the Gulf of Mexico. Increasing water absorption and holding capacity will grow bigger crops and reduce flooding. How wonderful it would be if the Mississippi was no longer the "Big Muddy" due to all the eroded soil that travels down it. We need to be building soil instead of losing it. Water that is leaving a field should be clear!

Also, paying farmers to sequester carbon, especially with biochar, will change practices. Changing corn acreage to more year-round growth to keep the soil alive while making biochar out of crop residue would be more beneficial and less wasteful for our environment. This change, while also concentrating on biogas digestion, will sequester nearly a billion tons of carbon through soil sequestration and biochar production. This is the amount given off by all the vehicles in the U.S.

Imagine if farmers, having applied large amounts of biochar to their land, no longer saw half of the fertilizer they bought wash away but remain on the land. Fertilizer companies would see their sales drop substantially, but farmers' pocketbooks would benefit, as would humanity, by conserving those limited resources for posterity and with a lower "carbon footprint" for the fertilizer industry. Regenerative farmers sum up the issue well when they point out that mainstream agriculture uses "toxic rescue chemistry." You can discover more on this topic and others at https://www.acresusa.com.

If the goal of carbon sequestration efforts is to reduce CO_2, one thing that must be done is to minimize carbon emissions while also sequestering carbon. The job of remineralizing the soil and making biochar is energy-intensive, and every effort should be made to do this work in a renewable manner. This proposal is bold, but it offers a new capability to do something positive with America's natural resources. There is a need for urgent action on a massive scale to prevent the steady descent to extinction through loss of soil and creation of permanent desert. Doing something positive about soil problems to also address the issue of rising atmospheric CO_2 is a bold but necessary step, as so many in the field don't give the soil much credit for its ability to sequester carbon!

Atmospheric CO_2 levels have had an "ebb and flow" cyclic effect of highs and lows throughout Earth's history with or without man's contributions to it. But depleted soil from over-farming can neither remineralize itself nor sequester carbon as it once did. It is essential to restore its health to pre-industrialized non-chemical farming levels.

A comprehensive approach to biochar may appear costly, but there is so much to gain even beyond the carbon benefits from a program focused on remineralization. For instance, by utilizing the deadwood and other new approaches to managing the forests, which is discussed in another chapter, forest fires will be dramatically reduced in frequency and scope. These fires are major carbon producers and very costly themselves, including the cost of lives and livelihood for some. The approach also has collateral benefits by encouraging greater growth in the forest.

This portion of The Get Real Program alone is a big win for our environment and quality of life, without making painful adjustments to the way of life based on plentiful and reliable energy.

CHAPTER 5

THE MIRACLE: SOIL LIFE

It is time to get our hands dirty and dig into our natural resource of soil. Soil is critical to life on this Earth – not only life itself, but also the quality of life. It plays a key part in supporting our lives through its critical role in the Earth's ecosystem. Our soil is deteriorating in many, many areas in an unsustainable trend. It is being depleted of required nutrients to sustain life and secure our food supply. The Get Real Program sees the future and is sounding the alarm, while also bringing forward solutions to solve the soil crisis.

Life will not exist as we know it if we don't heavily invest in our life-sustaining soil. The crisis is real! Without change, within our children's lifetimes, man will be largely starving or living on manufactured food. Our bodies are not made and have not evolved to be maintained or prosper on fake nutrition. Yes, we can get by for a while on depleted nutrients, but not for long and not with the higher quality of life we are used to experiencing.

We thrive best by going back to the basics with our nutrition and eating whole foods created by nature, not processed foods and beverages. So called "food advances" have only caused an increase in disease, including cardiovascular disease, diabetes, obesity, cancers, and more. Modernization has not "advanced" society in regard to nutrition. It is nature that can solve our food crisis and disease advances.

Although the soil crisis does not garner as much attention as climate change, the crisis we are encountering with our depleted soils is far more dangerous to our survival than climate change. That being said, soil is a big part of the answer to climate woes from rising carbon levels in the atmosphere. Soil is an essential "element" that is critical to life on land, yet it is in a nutrient deficiency crisis. Soil cannot help us unless we help the soil to thrive at the levels of the past. There are solutions!

Both climate and soil concerns are connected and can be solved simultaneously by revitalizing the soil. As the soil is revitalized by working within The Get Real Program, the rebuilding of carbon levels in the soil will help manage the level of carbon in the atmosphere. It is a win-win, no matter where you stand on the climate concerns. By re-focusing our attention on the soil, we can make huge strides today for a better and healthier tomorrow!

Soil can be one of Earth's most complex places, dense with life, or it can be a nearly sterile wasteland that is devoid of life. About a million acres a year are trending out of fertility and toward becoming desert, but we grow increasingly nutrient deficient food in increasing quantities to keep the world fed, albeit poorly, by some measures. Meanwhile, there is so much waste, and investment focused on the wrong areas of Earth's sustainability, rather than proven ways of securing a healthy and prosperous life for generations to come.

It is alarming to see the trends of neglect, negligence, and flat-out ignorance! Every single day, tons of vital nutrients are washed into the ocean or buried in a landfill. We need to return the minerals in human and food waste back into the land in an environmentally safe way. Without change, man is doomed long before the speculated changes from rising CO2 levels happen... if we don't save and rebuild the soil. There are many solutions that can create change and revive our depleting soils, including remineralization and regenerative agriculture.

THE SOIL AND CLIMATE CHANGE CONNECTION

Saving and revitalizing the soil can accomplish amazing things beyond the benefits to the soil itself discussed later in this book. One of these is the lowering of atmospheric carbon levels. The Get Real Program proposes a practical way to lower atmospheric carbon by bringing the soil back to life. Living soil requires carbon compounds derived from CO2 by photosynthesis for its countless life processes. The carbon content in the soil can build without end as long as there are rock element nutrients are available.

By investing in remineralization of our soil with biochar and rock dust, we are investing in sustainability of life throughout our Earth. Living soil depends on near-continuous photosynthesis occurring when it is warm enough for plant and soil life to flourish. In colder regions of the world, soil and plant life take a long winter holiday. Yet, these regions are the most mineral-rich as they are often in areas of previous glaciation, both ancient and more recent. It is soil in warm areas that never freeze and have heavy rainfall that is badly demineralized. This is generally the case unless there are other sources of minerals nearby, like volcanoes.

A BRIEF HISTORY OF SOIL

Some scientists focus on ancient soils, such as those in Africa, but the most fertile soils are often relatively young soils. Examples of young, fertile soil can be found in the Northern American Midwest. These soils were created after the end of the last ice age 10,000 years ago. When dealing with the ancient soils of Africa, 10,000 years is a short time period.

Some of the African soils are millions of years old, a fact known because of the prevalence of ancient fossils that can be found near the surface of the soil. The fertile North American soils from the geologically recent glacial past are a wonder, as they are imbued with rock and rock powders. The soil is created when a glacier grinds down into bedrock and leaves behind ground rock of varying sizes, from dust to boulders. The most fertile soil comes from glaciers grinding rocks into rock dust; however, even these lands are demineralized compared to a period called the Climate Optimum, around 5,000 years after the glaciers receded. This length of time allowed soil life to flourish and produce a fertile, rich soil that still had many unextracted available rock nutrients.

With plant life so abundant and soils so rich in organic matter, soil erosion was minimal, and the soil depth increased each year – instead of shrinking as it does in many places now. One reason that our current geologic time has so much life is that glaciers have produced a large amount of rock dust over the last two million years.

As glaciers grind rocks into rock dust along their slow travels, eventually, some of the rock dust comes out from under the glacier. When this happens, alluvial fans are formed in periods of glacial growth and retreat. Summer winds created by large temperature differentials near the ice pack carried this fine rock dust as far as the equator, remineralizing soils worldwide. Of course, non-glacial areas were where more of the plants and animals lived. Over millennia, ecosystems consumed rock dust, leaving behind tiny clay particles that wash into the sea.

Soil life has become less abundant as available rock nutrients have depleted during the long, current interglacial period. Some scientists feel rising atmospheric CO_2 levels have broken the glacial cycle, increasing greenhouse gases to a point where the remaining areas of ice will shrink away rather than regrowing. While scientists speculate as to what causes the approximately 100,000-year glacial cycle, they tend to agree that another glacial advancement may be due since the Earth is so demineralized.

This is where an intervention by man is critical to support continued soil sustenance. Fortunately, people can remineralize the earth in a more efficient, less disruptive way than glaciers can. People using energy wisely can grind the abundant basalt rocks and spread the dust on land around the world.

Basalt rock has all the elements needed for microbial life, proven to boost both soil and plants, and its nutrients will not leach away and are in place until extracted by soil microbes. And we can help Mother Nature along with a little, or perhaps a lot of, human innovation. Sometimes man is given the tools to create the answer, like with our soil, and here we can work within nature to nurture our soil for a healthier way of life.

MICROORGANISMS AND OUR SOIL

By weight, the vast majority of life on land is comprised of microorganisms in the soil, and all other life relies on productive soil for sustenance in one way or another. How microorganisms reproduce and spread is only partially understood. Since they are invisible without an optical microscope, our knowledge is based on conjecture and theory rather than direct observation.

Microorganisms, and organisms like lichen, also live on rock, converting its minerals into complex compounds containing carbon from the air. This eventually produces a porous and nutrient-rich growing medium for plants to emerge. The presence of plants further increases the soil's organic matter as sunlight converts atmospheric carbon into complex carbohydrates and proteins through the process of photosynthesis.

There are two types of organisms in the soil. One we can call the composers, which are agents that use rock nutrients to make previously unavailable nutrients available to plants. Many composers turn carbohydrates from photosynthesis into stable carbon compounds such as components in the mysterious humus. The other agent we'll call the decomposers, which attack organic compounds and things like plants and release carbon back to the air as $CO2$. There is a constant battle between the two agents. Some creatures, like earthworms, are both composers and decomposers.

At its core, healthy soil is made up of rock of all sizes and types. Organisms eat certain types of rock and create a variety of complex compounds that make up healthy soil. However, some forms of rock are devoid of essential nutrients. An obvious example is sand (silicone dioxide or SiO2). It is one of the most common compounds on Earth, but it's sterile and unable to support life in its pure form. At the other end of the fertility rock spectrum is basalt, which has a very wide variety of life-supporting nutrients.

There are vast areas of the world that don't support life. These areas are growing along with the demands that come with a larger population. Something needs to be done before we have to make tough decisions like rationing and limiting family size.

THE CHEMISTRY WITHIN

Soil becomes increasingly acidic as rock nutrients leach out with rainfall. Soil that is either too acidic or too basic does not allow plants to grow well. Soil life does not thrive in acidic (low pH) soil, as that means there are few available rock nutrients. As important as biology might be for healthy soil, chemistry is key. For example, some elements are unavailable at either low or high pH levels due to chemical reactivity.

Calcium and magnesium go "hand-in-hand," says soil and fertilizer specialist Neal Kinsey, owner of Missouri-based Kinsey Agricultural Services, Inc. In his book *Hands On Agronomy*. "Magnesium, in conjunction with calcium, is the key to air and water in the soil," says Kinsey. The so-called "cation-exchange capacity of soil" measures the soil's ability to hold positive nutrient ions, such as calcium and magnesium, and components such as clay particles and humus play an important role.[7]

Ideally, soil cations are attached to soil nutrient ions for plant availability. Dr. Albrecht was a pioneer in determining the ideal

[7] Kinsey, Neal and Walters, Charles, ***Neal Kinsey's Hands-On Agronomy: Understanding Soil Fertility & Fertilizer Use***, Acres U.S.A., Greeley, CO, 2013.

amounts of nutrients in the soil and observed that with 68 percent calcium ions and 12 percent magnesium (with lesser amounts of phosphate and potassium) present, optimal growth occurs. Sadly, most of the world's soils are so terribly demineralized that they simply do not have anywhere near these amounts.

Albrecht was a true pioneer in soil science, yet he suffered professional criticism for his groundbreaking theories on what made soils, plants, and ultimately animals healthy. He diverged from many of his peers by not embracing the big three nutrients (nitrogen, potassium, and phosphorous) as the "end-all" of nutrition. Rather than just looking at soil acidity levels (pH) as the primary factor to be controlled, Albrecht looked at the importance of having a proper ratio of calcium and magnesium. Calcium and magnesium are generally mined, ground, and sold from local quarries at low prices, so there is only limited commercial incentive to promote more calcium and magnesium use.

After a long period of denial by many in soil science, Dr. Albrecht's principle of the optimal ratio of calcium and magnesium, defined as percentage of saturation of the soil's ion exchange capacity, is now widely used. Yet, most farmers still don't work to get calcium and magnesium in balance, and much of the world is very badly demineralized. In fact, much of the world's soil is now in desert, a state totally out of balance and almost completely devoid of life.

When you have a world in which a large portion of the non-frozen land is either a desert or on its way to becoming one, it is clear that there is a crisis that cannot be ignored. If you want to read more on this topic, books and lectures by Dr. Albrecht's student Neil Kinsey can be found at Acres U.S.A Bookstore. The bookstore also includes Dr. Albrecht's classic works and much more. Go to https://www.acresusa.com/ to get lots of resources [8]and sign up for their magazine.

---

8 https://www.acresusa.com

THE SOIL AND YOUR HEALTH

Each day, tons of vital essential nutrients such as phosphorous and potassium are washed into the ocean or buried in a landfill while we dig up concentrated mineral deposits to make up for that misuse. Sometimes a majority of the mined minerals placed on farmland just washes away into the sea while a fraction feeds an unhealthy plant that is deficient in the other numerous nutrients that are not in the chemical fertilizer.

Is it not obvious that often such food will be unhealthy and result in health problems for those who eat it? We must quit wasting nonrenewable, depleting resources out of laziness and greed. The miners of minerals want to sell more now, rather than stretch out the supply for centuries. We can't concentrate most of the world's population away from growing areas and just waste all the nutrients by trucking crop residuals and weeds off the ranches and farms into the landfills or sending runoff to the sea. Also, we need to return the minerals in human and food waste back to the land in an environmentally safe way.

Nutritional Value

Bethany Davis, director of MegaFood and its related website/blog www.*MegaFood.com & MegaBlog*, puts it bluntly: "We have 60 harvests, or fewer, left. Not only is our food less nutritious than it was 50 years ago, but we are destroying our soils at such a rate that we won't be able to feed the world." She continues, "The fact that our soils are collapsing is what is driving our focus on regenerative agriculture."[9]

Man tends to fixate on quantity of production rather than quality of produce. Anyone can identify a big ear of corn, but what is more difficult to determine is the quality of an ear of corn. Ultimately,

---

[9] https://www.megafood.com

the real sign of quality produce is not the measurement of protein or energy within, but how well it feeds a successor animal. Unfortunately, because most soil science is based on simple measures of improvement, such as yield, plant breeding has been focused almost entirely on quantity, often at the cost of a declining level of quality.

A striking example of this can be found in apples. Today, apples yield more tons of fruit per acre than at any time in recorded history, yet the nutritional value per ounce of apple has plummeted. One would have to eat three to four of today's apples to get the same level of a certain nutrient gained from eating one apple in 1965! This striking drop in quality and availability of certain nutrients is a result of years of wind erosion, rainfall, and crop harvests steadily "mining" the soil as well as the push to faster and bigger growing insect resistant fruits and vegetables. The problem is that many of the resulting fruits and vegetables have less nutritional value per ounce for the consumer. A 2011 *Scientific American* article stated that the nutritional drop was significant and reliable on average for up to 43 nutrients in a respected university sponsored study they reviewed. Other studies cited in the article found drops of 15-30% in certain nutrient content.

With about a million acres a year running out of fertility and becoming desert, we are continuing to grow nutrient-deficient food in increasing quantities. This gives us something to put in our mouths, but it is not giving us what we need to be healthy and prosperous. One of man's strengths is his ability to adapt and innovate. When the soil got increasingly depleted, we bred plants that could grow big yields in depleted soil.

We told ourselves that white bread was good to eat and that falling levels of protein and minerals in the flour were of no consequence. Wheat yields in bushels have risen, but the minerals and protein have dropped sharply. There is a fabulous book called *Bread from Stones* by Julius Hensel, which investigates these issues in great detail. These issues can be addressed with a change in attitude and focus.[10]

People and animals crave quality nutrients in food. If the food doesn't have them, they eat as much as possible to try and gain them. Cows and humans equate sweetness with quality because in a natural environment, nutritious food is sweet, and eating good-tasting food leads to good nutrition. Eating a fat animal in the wild means it has had plenty to eat and has good minerals and vitamins that often are in the fat. Another example is that fat from grass-fed cattle is more nutritious than fat from feedlot corn-fed cattle.

We are being fooled into malnutrition by artificial sweetness and unhealthy fats. These complications are results of deficient soil. Deficient soil will grow deficient plants, which will produce unhealthy humans and animals. William Albrecht (1888-1974), chairman of the Department of Soils at the University of Missouri, summarized the issue nicely: "When soils are more fertile and higher in mineral fertility, their crops are those normally richer in the nutrient elements like calcium, phosphorus, and other minerals."

People and plants can live on very unbalanced and ultimately unhealthy diets for periods of time, but will no doubt suffer from disease as a result. One of Albrecht's very clever studies involved feeding trials of plants grown with different fertility regimes. He grew crops and fed them to rabbits. He then observed and weighed the rabbits fed identical plants that were fertilized differently. He found that the healthiest, biggest rabbits came from eating well fertilized crops that had their minerals in proper balance. These crops might not have been the biggest, but they had far more nutrients in them.

A rabbit can only eat so much, so it is important the food be as nutritious as possible. People too can only eat so much. Most people would never consider eating three of today's apples to make up for the missing

[10] Hensel, Julius, ***Bread from Stones***, Greeley, CO, Acres U.S.A., 1991

nutrients, when, in 1965, one apple would do. Today, many people subsist on a demineralized diet, whether they are aware of it or not.

Organic Is Not the Solution for All

Even those trying to do the right things by eating organically are finding nutrition deficiencies. Unhealthy eating can lead to low energy, foggy thinking, underlying disease, and more problems. We must invest in our soil as a building block for better foods and nutrition on top of the benefits to our atmosphere.

Some organic farmers and advocates don't do comprehensive soil testing and rock mineral application as a main part of their program. The market is full of organic produce that is not mineral-rich. All "organic" means is that manufactured fertilizer and pesticides are not used in food production. A labeling protocol to designate food produced from fully mineralized soil is necessary.

MAN'S ROLE IN OUR SOIL

Soil can perform much differently than it has under man's mismanagement. Man's practices have played a key role in where we are today with our soil. Innovation is great, but in some cases, we have to get back to basics and work with Mother Nature and not try to work around it. Soil is an asset that can produce so much for us, if we just value it for what it can provide, instead of stripping it of resources and not reinvesting in it through remineralization. From a business perspective, mining the soil is like mining for minerals. One can work until the mine no longer has anything of value. The miners are in business until the mine is barren.

Chemical Fertilizers and Processes

A sad turning point in the understanding of the soil came with a chemist named Justus Von Liebig. Von Liebig isolated some of the primary elements used by plants and produced them chemically.

He saw increased growth from applying the chemical nitrogen and potassium to the test soil. This discovery marked the beginning of the chemical fertilizer age. For decades, mainstream soil science focused on the manufactured fertilizers and ignored other essential elements for healthy plant growth. The resulting plants were imbalanced and invited insects and other parasites that attack sick plants.

The difference in soil life between regenerative agriculture and the dominant chemical farming systems is like the difference between the soil life of a lush garden and that of a desert. An earthworm will rarely be found in a chemically farmed field but can number in the hundreds in a cubic yard of properly managed soil. The problems are deeper than what we can see... mostly down in the roots.

The chemical crowd thinks that roots are simple, one-way streets for water-soluble nutrients, while they are, in fact, much more complex, especially in a regenerative system that boosts soil life. Some farmers even inoculate their fields with extra microorganisms and enzymes to restore sick soil. There is always a give-and-take with nature and anything else worthwhile. The Get Real Program knows there is a cost for proper care and attention to our soil for a healthier life and environment, so it is important to give support to the farmers. Farmers, in turn, need to give more support to our soil, so that we get more nutritionally abundant crops for our long-term prosperity, sustenance, and longevity.

The financial pressure of managing a large property too often leads producers to make decisions for short-term survival. As soil improvement is a long-term gain, it is often neglected, and focus is directed towards the monthly or annual bills. We need a system that encourages farmers to rebuild and save the soil, and this book proposes a viable solution. One thing we all need to do is avoid being cheap when it comes to food prices. As a society our attempt to save a few dollars drives producers and workers to low margins and even poverty. When possible, buy premium organic brands and shop at locally supplied meat and food shops or markets!

Innovation, Genetic Advancements, or Depletion?

Short-term monetary incentives play a role in the mining of soil. There are scholars, agronomists, and growers who know better ways to produce nutritious food without chemicals, but one can make more money mining the soil in a short timeframe than farming sustainably. This is especially true if the compensation for food is based on quantity rather than nutritional value. Malnutrition, obesity, diabetes, and other ills are the results.

With automation came a lack of respect for the land and the adoption of chemical farming with manufactured and often toxic compounds. Additional changes have come to agriculture in recent decades. Plant breeding has changed from a farmer selecting the best plant to keep for seed on his farm to an incredibly technical enterprise.

Today, it involves things like taking genetic material from one species and inserting it into another plant's genetic material. The results of such an operation can achieve ends like a bio-derived insecticide from bacteria being produced in a corn plant. Now the seed cost is enormous, and pests are adapting, making the GMO seed ineffective in just a few years. The solution is regenerative farming and restoring the soil to health to produce healthy plants that are naturally more insect resistant.

The Get Real Program understands the need for innovation, automation, and creativity in solving our soil crisis. It is important, however, to not create a new crisis by trying to solve for the depleting soils. We have learned much by the impacts of genetically modified corn and other quick fix practices that should be addressed in how we go about our farming and caring for our precious resources.

We're at the end of an interglacial period and the soil is at a peak demineralized level. It is no wonder that the world is seeing increased obesity, diabetes, and other severe illnesses. All of the innovations may be hurting us as much as helping. The Get Real Program has answers and solutions to dramatically change the trajectory of more than the rising CO2 concerns. It has REAL solutions!

Building Out of America

Desertification through the destruction of fertile land by man's non-stop land grabs and building is adding to the concerns over Earth's sustainability. Man cannot cover all of our earthly surfaces with manmade structures and cement. Unfortunately, man tends to build on nearly flat ground that often has topsoil instead of choosing the hilly eroded land for housing. Each year, millions of acres of productive ground are covered with concrete and buildings. That land is effectively lost to potential soil carbon sequestration and agricultural productivity.

For too long, we have treated the land as if it were without limits, and we just intensify production on the remaining acres. We have benefited from the long warming period of the last hundred years to move crop production north. We should no longer build on fertile land, but, instead, use only low-grade agricultural land for housing and business. For too many years, man has covered fertile land with structures without regard for having enough land to properly feed us. It is not just the building that is a challenge, but also centralizing the settling of man.

An unfortunate trend in agriculture is the population of workers moving to the city. This has been a trend for over 100 years as mechanization came to the countryside and dramatically reduced the amount of labor needed to work the land. To achieve sustainability, the concentration of wealth in cities and the gradual stripping of resources and people from the countryside must end. We live in a country where our food is grown by very few people and transported and stored for lengths of time great distances from where it is grown to be consumed. The time it takes to move them contributes to the depletion of nutrients in our foods.

The Get Real Program is very interested and supports advances and innovation, but not ones that harm our resources in a way that puts our existence at risk. The advances must work to create more and give back to our natural resources in a way that creates advances for humans in the long-term, not just empty, short-term gains. Let's work together to create a more prosperous future with the solutions laid out in The Program.

VALUE OF PROPERLY INVESTING IN OUR SOIL

Many forms of wildlife, such as deer, are regulated by the government to ensure their survival. Soil life, however, does not have any protection despite its critical role in the long-term healthy survival of humanity. Many scientists treat the soil as a passive exchange medium to hold vital nutrients and provide a base for plant life. Yet, a spoonful of healthy soil is itself home to a billion life forms ranging from enzymes to earthworms. Sadly, the process of degradation of nature and our soil by natural and manmade processes is on the rise. The world faces a shrinking pool of fertile soil and a loss each year of marginal soil, which has little agricultural value as is. There are many areas of the world where one can go hundreds of miles without encountering fertile soil. Frankly, due to man's practices there is limited truly living soil to be found.

Despite these concerning trends, there is reason for optimism. Soil life is capable of making a comeback when the conditions are once again right for it to flourish. Like seeds, soil life may lie in the ground for generations, awaiting the right conditions to come about for regeneration.

YES, THERE ARE SOLUTIONS... BY THE GET REAL PROGRAM

Restoration of Rock Minerals

As mentioned, restoring rock minerals is a proven way to sustainably improve the soil and crops. Too often, scientists don't look at difficult-to-duplicate biologic processes carried out by soil organisms that are so easily killed by man or made to go dormant. As a farmer, the founder has seen restoration of rock minerals create value when growing massive, nutritious crops for his cattle to eat by restoring most soil nutrient levels.

A lack of affordable rock dust containing a wide array of nutrients led me to use commercially available micronutrients as needed. There would be greatly reduced demand for fertilizer products if it was widely known that simple rock nutrients can revitalize soil and grow nutritious plants. Yet, as remineralization is not widely practiced, large profits are made from toxic proprietary products and concentrated fertilizers.

The harsh truth is that research requires funding, and most agricultural researchers follow the money. Inexpensive and widely available rock dust is currently promoted only by a very small charity and a few business people making a small effort to encourage remineralization. Please go to www.remineralize.org to learn more and donate to make a real difference in soil health. (A portion of any profits from this book go to support this group.)

Cattle Connection

While many believe a big reduction in cattle is desirable for curbing climate change, the Savory Institute advocates an opposite approach. Allan Savory, of the eponymous Institute, has found that by grazing cattle holistically, grasslands can be regenerated, and total net carbon will be sequestered in the land. How can there be two positions so opposed? The anti-cattle advocates misunderstand the full functioning of a grassland, often giving grass no credit for sequestering carbon.

Savory is an expert in changing near-desert land back into productive grassland. He stresses the need for more grazing by cows in a method that simulates the natural grazing habits of wild ruminants under pack predator pressure. Constant movement with intense, very short-term grazing followed by a substantial period of rest, is the natural cycle that causes grasslands to flourish.

It is easy to forget how different the western United States was 200 years ago. The presence of predators was nature's way of controlling the grazing animals that kept the grasses vital. The elimination of the wolf so that domestic livestock could be raised has resulted in degradation of the grasslands to the point where there is very little grass left on millions of acres. The only vegetation that is present is inedible. Traditional forms of grazing or the lack of grazing altogether results in the desertification of land. Only a return to a form of grazing similar to the way it existed centuries ago will reverse devastating desertification.

In addition to the necessary changes to cattle grazing, another potentially destructive animal must be dealt with – the horse. The horse, introduced by the Spanish, has bred to numbers far in excess of the ability of the land to sustain them for a long time. They have become a desert-causing creature. Horses graze grass to the ground and are the cause of much of the loss of the native American prairie grass on the land that wasn't plowed into oblivion by the early settlers. The free roaming of horses on federal land must end to help prevent the desertification that is well underway (see free auctions of mustangs on Federal Land).

Much of the land in the American West is owned by the federal government. There are many interest groups that want to use the land and many others that want the land to be left unaltered.

Unfortunately, the Bureau of Land Management (BLM) does not require positive holistic grazing. Instead, the land is leased out for traditional grazing, which destroys the native grasses and degrades the soil. The BLM doesn't do any soil tests or improve the land in any way, allowing it to steadily degrade.

Savory's excellent book *Holistic Management* has this to say about the soil:

> "Any changes brought about above ground are likely to cause even greater changes underground, simply because there is generally more life underground than above ground. Figures vary widely with different soils, but on average, the top six inches of soil of grassland soils have 7.75 tons of microorganisms per acre such as bacteria, fungi, earthworms, mites, nematodes, and protozoa. Now, that is moderately healthy soil. The richest soils can have up to 15 tons per acre while chemically farmed ground may have very little healthy life."

The ideas of Allan Savory and the Savory Institute need to be utilized to formulate policy to transform much of the near-desert of the southwestern United States into carbon sequestering areas.

Overall, traditional grazing and the wild horse have profoundly damaged the grasslands of the American West. Without grass to protect the soil, it oxidizes and erodes. One of the problems with the desertification of soil is that it loses its organic matter and becomes hardened so that water can't easily soak in. This results in rainwater quickly running off and producing damaging and eroding flash floods.

Restoring native grasses to areas that are becoming deserts will change that environment and make it able to sustainably support regenerative grazing. Some animals that are adapted to live in near-desert conditions will die out as the environment changes to be more productive. See Allan Savory's website or www.soilcarbonunderground.com for vivid evidence that we really can change the world for the better.

Biochar

There are actually people who would say that we should leave deserts alone (some are afraid of damaging or changing the ecosystems as they are today) or not try to prevent land from changing to desert. Our goal is the opposite: to restore the Earth to a higher state of health and provide plenty of food and resources for people while also protecting wildlife. Yes, there will be changes to the planet, but they should be positive changes for increased sustainability.

As discussed previously, making and placing biochar consistently sequesters carbon for centuries, and creates an environment in the soil where soil life can flourish and sequester far more carbon in a stable, long-lasting form. Putting out massive amounts of biochar will sequester water-soluble nutrients on the land. This will lead to a sharp decline in toxic red algae in the coastal areas.

A substantial commitment to charities focused on carbon sequestration will allow funds to be directed to encourage farmers to change their ways from being carbon emitters to sequestering billions of tons of carbon through remineralization, use of biochar, and better practices such as year-round growing to feed the soil. Biochar production, usage, and benefits are discussed more in other chapters within this book.

Simple, Clean Farming Is Key to the Solution of Rising Carbon Levels

Farming is a solution for many issues that we are facing in our environment, including rising CO2 levels, and the health of Earth generally today. To succeed in turning the world's farms and ranches into active carbon sequestration sites while lowering CO_2 levels, we need to take reasonable actions. The changes being discussed will have added positives beyond carbon. They will also increase the nutrition of food and increase profitability of the farming properties. Many books could be written on the solutions of farming and caring for our soil, but the focus of many is on the rising levels of carbon instead.

Conventional crop farming is a net carbon emitter as most of the ground is bare and oxidizing, even during the short growing season. In northern regions, the soil freezes. Life is essentially put on hold so there is no carbon oxidation. They have cover crops growing up until the freeze, and they are ready to spring back to life as soon as the ground and plants thaw out. In areas where the ground doesn't freeze, some plant life can continue all winter. The traditional mindset wants bare ground when a crop isn't growing. Yet the result is destruction of soil carbon as hungry soil microorganisms feed on it. Tilling of soil adds oxygen to further burn up soil organic matter. Many temperate and tropical soils are already badly demineralized. They can't support humus-building microorganisms and don't have healthy fungi or other microbial life in the soil.

In many cases, these areas have thin layers of topsoil if they have topsoil at all. Under topsoil, too many areas have only dead subsoil insufficient to support plant life. It takes very little organic matter to support plant life, but without any, nothing will grow. As a farmer, the Founder had success revegetating bare, sterile subsoil with only a very thin layer of mulch and manure. Again, working within nature and not against it, is the answer to replenishment of our soils and nutritional values of the soils and ultimately humans through our food supply.

At the Savory Institute, there are amazing pictures and case studies of millions of acres of land restored to vitality by holistic management of the soil! The Get Real Program promotes a radical, but sensible, and proven effective shift towards regenerative agriculture. There will be winners and losers in this change, but the overall winners will be Earth and humans as we see prosperity grow from our soil and feel the impacts of the increased nutrition. Some very profitable companies will see their revenue fall sharply as farmers use fewer chemicals, while many other businesses with positive long-term thinking and processes will thrive.

What could food production look like in the future with a more sustainable system? For one, agriculture will make sequestering carbon in the soil and producing biofuels a large part of a farm's operation. The current system utilizes a "just-in-time," no stockpile system where there is only a small amount of extra food, especially when it comes to meat. Another problem with the current system involves exporting grain. When grain is exported, its nutrients leave the farm, never to return. Making it worse, soil loss is many times the weight of the grain leaving the farm due to bare soil and the wind and water erosion.

The Mississippi River sends many millions of tons of vital topsoil to the ocean each year. We even call it Big Muddy. This folksy name relays the magnitude of the problem. The massive levels of erosion we witness should not be the status quo. Fully covered soil sees little to no erosion. Truly healthy mineral-rich soil grows organic matter each year, getting richer and more productive over time. A regenerative farmer uses no poisons that kill soil life and, unlike a conventional farmer, sees his land get better over time.

The book *The No-Till Organic Vegetable Farm* by Daniel Mays nicely summarizes the practice:

> "Organic no-till depends on cover cropping for weed management, but it can also help substantially with nutrient management. Your crops' nitrogen needs can be addressed through the inclusion of legumes in your rotation. And because the soil's organic matter stores nutrients in a bio-available form, you

> will be building a store of nutrients that will last for years to come."[11]

The book is available at the *Acres USA Bookstore* and has lots of good information about how to enact no-till without the toxic chemicals most no-till farmers use.

One way in which organic no-till farming is superior to Inorganic no-till farming is that with organic no-till, crimpers are used to kill the cover crop. This forces part of the crop into the soil, where it can decompose into soil organic matter, whereas with inorganic no-till farming, the cover crop remains on the surface and does not benefit the soil. Some in the alternative agriculture space don't embrace soil testing and adding minerals to the soil, but organic produce is often mineral-deficient, and there is a need for the adoption of broad-spectrum remineralization. The first step is to restore soil life, and that means not poisoning it but feeding it with continuous living plants and minerals.

We cannot forget permaculture as a solution either. Permaculture is a growing international movement that offers a different way of looking at the land and how we create value from it. Permaculture aims to create a holistic, multi-species, nearly permanent environment that sequesters nutrients and water on the land. The practice of permaculture is essential for long-term sustainability. Permaculture is on the right track, using plant growth to build healthy soil and creating organic matter through composting and mulching. It can be seen as the opposite to the traditional but very destructive clean tillage monoculture. Quality information about permaculture can be found on the internet in addition to the many good books on the subject. See www.permaculturenews.org for more information.

---

[11] Mays, Daniel, ***The No-Till Organic Vegetable Farm: How to Start and Run a Profitable Market Garden That Builds Health in Soil, Crops, and Communities***, North Adams, MA, Storey Publishing, 2020.

Earthworms to the Rescue

One important indicator of healthy soil is the presence of earthworms. They are indicators of healthy soil because they require organic matter to live. Additionally, they do a number of things to make soil life better, including aerating the soil and moving vital nutrients where plants can access them. Sadly, earthworms are highly sensitive to the chemical balance in the soil and are badly affected by modern chemical farming.

In the modern agricultural field, it can be very difficult to find a single earthworm, whereas in the rare fertile soil, they can be found in abundance. Instead, much American farmland is blighted with parasites such as nematodes and root worms due to sick monocultural farming practices. We need to have a national goal of hundreds of earthworms per square yard of soil.

Conventional farmers respond to parasitic threats with toxic rescue chemistry that kills more soil life. The worst offender is anhydrous ammonia. This product is still used by many as a source of nitrogen for crops, despite its dangers. It is used because it is a cheap, highly concentrated form of nitrogen. It needs to be banned because it is so destructive to soil life and organic matter.

Most people over 50 years of age can all remember the earthworms in our youth, but do not see them as often today. We need a comeback of the earthworm. Whatever you thought as a child about the value and meaning behind your encounter with the earthworm, we bet you will look at it differently the next time you see one. You will look at it as a marvel of nature and something of value, rather than a toy or something gross.

Carbon Sequestration

The Get Real Program is full of solutions for the health of the planet and the health of the reader of this material, and carbon sequestration is key to creating real change. There is detailed information shared on this critical solution in later chapters. There are many solutions for carbon sequestration, but the soil is a major cog in the larger wheel of survival. A soil carbon sequestration program must include universal soil testing for minerals and soil carbon content. Ideally, differing rates of compensation will be paid based on how long carbon is removed from the atmosphere. There is more on carbon payment mechanisms laid out later in this book.

Estimates of soil carbon sequestration from many mainstream scientists range from only five percent to 10 percent of what healthy soil can actually do. The actual big potential is hard for scientists to mimic in a lab setting and is only being learned through books and lectures. The founder has seen the power of healthy soil, which is a magnitude above the mainstream estimates!

THE TIME TO BEGIN IS NOW...

As an example, the effective practice of making small holes in hard desert soil and putting a little organic matter and a seed in to create a micro-environment for a plant to grow could be improved with the use of nutrient-laden biochar and rock dust. By adding these materials to water catchments, photosynthesis and carbon sequestration can be stimulated in lands that are currently carbon emitters.

Charities funded for real impact could contract young generations and other labor groups to do public service regenerating the desolate regions of the western United States. Such an effort could avert disaster in a region that is becoming a larger desert due to bad policies and neglect. It could also provide meaningful work for physically able, non-violent criminal offenders.

Once native grasses are established, the land can be holistically grazed using the Savory method. This method will encourage grass to spread and eventually return the area to fully covered grazing land. Understandably, some areas may have too little soil to become grassland and may require other types of plants in the water gathering areas that are loaded with biochar and rock dust.

TRUST IN THE FARMER, NOT THOSE WITH LINED POCKETS OR THE RESEARCHER IN A LAB

Almost all research soil is blended, a process that shatters the organisms and sterilizes the soil. Because of this, before research is done, most organisms are killed either accidentally or on purpose. Testing dead soil for its properties is like testing a dead animal. This is a problem that needs to be addressed in order to conduct better research. Plants have amazing properties to be able to function to some extent in very unhealthy soil, but the resulting plants are not really healthy. Conducting research on healthy living soil can result in more nutritious successful plants, but many times a scientist has not actually worked in the natural soils to know what truly healthy soil looks like.

Now you might ask by what authority we speak about the soil and dispute learned men. The answer is decades of on-farm work along with extensive reading on the subjects and many discussions with independent regenerative proponents. Additionally, reading the conventional wisdom led to a broad understanding of both the chemical and natural agricultural processes.

The founder grew up in the city helping with a large garden; he comes from a family with a legacy of success. Seeing the depleting countryside and growing population as an opportunity, with his family the founder bought a large farm, which he operated. He read widely and learned a lot through decades of experience. The result was a successful grass-fed natural beef business. This experience, as well as many other experiences on the farm and in the industry, led him to want to create and innovate for better quality lifestyles and the environment. Every day is an opportunity to create a healthier world and he loves being part of the solution through the efforts of The Get Real Alliance and its Program.

IMAGINE A HEALTHIER, MORE PROSPEROUS WORLD FOR ALL

The physical loss of topsoil is an existential threat, and we need to sharply reduce the amount of soil that erodes away – ideally, creating a situation where soil is growing, not shrinking. If the soil is gaining organic carbon every year and you're feeding microbes that are breaking down rocks, you're producing new topsoil. This creates a net growth instead of a net shrinkage of quality soil.

Not only do we need to produce new soil, but we also need to care for the soil we already have. The nutritional need to balance nutrients and have calcium and magnesium in the proper ratio in the soil is really important. There are a lot of great resources out there that share more on the topic of our soil crisis, like www.acresusa.com.

Many soil scientists think it takes 1,000 years to make an inch of topsoil. Those who hold this belief have never worked with healthy living soil that is mineral-rich. In much less than a few thousand years, the Midwest of the United States went from crushed rock and mineral dust piles to several feet of fertile topsoil with added feet of subsoil below; many regenerative farmers have upgraded their soil substantially in just a few years.

The good news is that we can change things and really start sequestering carbon in the soil to an amazing degree with remineralization and sound farming practices. We can go from a satellite photo that shows desert covering most of the land mass to one with more green. It will take money and the right attitude to do so, but we have faith that America will do what is needed once we get this information out into the mainstream.

To create positive change and move forward in a healthier, more positive direction will require an investment and a movement away from the cheap practices of the present. We must compensate the stewards of the soil well for improving and restoring it. Because so much land in the American West is government-owned, there is a need for a change in the way the government manages it.

There are so many solutions that are beyond what we can share here, as well as much more material that continues to evolve on each topic within this Program. Therefore, please be a champion for healthy soil and join groups like www.remineralize.org and www.thecarbonunderground.org to network with others with similar interests and to spread the word. Our human survival depends on the health of the soil, literally. As mentioned earlier, the very comprehensive book, *Geotherapy*, has data on remineralization and other methods of improving soil and plant life. Together, we can make a difference and see a more prosperous future, but it will take inspired action to create change!

CHAPTER 6

THE MARVEL: GRASS AS A CARBON SOLUTION

Grass is more than just something nice to see or to sit on during a wonderful afternoon picnic. It can be part of a comprehensive solution for one of the most concerning issues facing the world today—rising CO2 levels—through carbon sequestration and more. As an experienced farmer and rancher, the founder has a unique vision to create real impact on the increasing CO2 levels all over the world by investing in holistic grazing and driving innovative solutions with prairie grass through creative inventions. Several destructive ways of man and flawed rhetoric on cows have misled the public towards imperfect, ineffective "so-called" solutions and away from natural holistic resolutions for real change.

Grass can lead the way on multiple fronts. With grass as part of a potential solution, just imagine the impact it can have, as it covers so much of our world. Like the ocean, grasslands can be put to work creating real impact according to The Get Real Program's ideas on holistic grazing and inventions to advance prairie grass to significantly improve health of our environment and food supply. There are other benefits to the solutions in The Program beyond the health of our environment. For instance, commercializing ideas for an integrated prairie grass operation can be much more fruitful for our economy, by bringing in a lot more money than a conventional grazing or crop (such as corn) operation.

Tell Me More About Grass…

Grass is an unusual plant. Unlike a tree, it is not designed to grow for a long time but to have a spurt of growth before becoming mature. It then starts over with new growth, leaving the old parts to die off and deteriorate. Doing so allows sunlight to reach new growth. If left on its own, grass will achieve less growth and can even die out in dry climates.

Surprisingly, fire can benefit grass. After a fire, grass can spring back from its roots and stockpile seeds that need an additional fire to activate. Yet grazing is preferable to burning when it comes to carbon. Random and intermittent grazing causes the grass to sequester more carbon in the soil, while fire releases an abundance of carbon into the air. Of course, unplanned and unmanaged fires can spark up, causing dire consequences and devastation. Therefore, it is better to manage our natural areas to prevent large wildfires while also moving grazing animals often, rather than leaving them alone.

Many activists look beyond grass to other solutions like tree planting. We will talk more about this in a later chapter, but while fond of tree planting as part of the solution, it is not a sufficient solution. Trees can definitely be useful in storing carbon when grown on healthy soil. Unfortunately, this solution has limits, as much of the Earth is too arid for trees to thrive.

The good news is that grass can grow and thrive in very arid climates if it is properly managed. Interestingly, grass that is not grazed tends to die out just as grass that is grazed too often does. A delicate balance allows grass to thrive. This is why it is critical to look to farmers and ranchers, rather than a scientist, for natural solutions to natural problems. Farmers and ranchers come from a place of experience with nature in the outdoors.

So much ground is nearly sterile due to mismanagement and demineralization that it needs to be inoculated with beneficial organisms to function well. This is where the cows, but not horses, come in…

THREATS TO OUR GRASS

Our grasslands need to be preserved and utilized for all of the benefits that they can offer the Earth, rather than destroyed through man's practices or ignored as they become sterile wasteland. Grasslands are part of the comprehensive solution of The Get Real Program. The list of threats and abuses to our grasslands is long. Below, we will mention a few that are most concerning to us, as a farmer/rancher and engineers who understands the critical role that grasslands play in the health of our environment and food supply.

The Dwindling Dung Beetle

The dung beetle plays a critical role by spreading out and burying cow manure, fertilizing the ground. The dung beetle forms small balls out of manure and rolls them to burrows as a future food source. Healthy pastures that host cattle will have an abundance of these curious creatures. Unfortunately, many pastures are rife with poisons to kill worms and parasites. Thus, the dung beetle is often absent from such environments, which plays a critical role in the sterilization of the land.

Nature knows how to care for itself. Many deficits we are seeing within nature and its fertilization efforts are in response to man's interference in natural processes. As hard as man tries to mimic nature, he cannot fertilize as well as nature is able to do. The more and more absent dung beetle is just one example of how man's interference impacts the Earth's natural ability to fertilize itself. These creatures support the life cycle by the critical role of spreading out the manure and burying it. This is the most opportune way the manure's nutrients can benefit plant life, instead of the unnatural process of oxidizing it or providing it on top of the ground. In the case of the dung beetle, man has interfered with nature's plan by inadvertently poisoning the beetles.

The Impact of Horses

Holistic grazing is part of the comprehensive solution within The Get Real Program, so a threat to grazing is a threat to the environment. Wild horses are not good for the land and will ultimately starve out after they have destroyed what little grass remains. In fact, the horse is one of the most destructive animals to holistic grazing. They were brought to America by the Spanish and then they multiplied quickly. Of course, they were critical to the early settlers, so people had large herds of horses. The challenge is that we are still dealing with the residual wild horses hammering the federal lands in a horrible way now that they are no longer needed for our day-to-day functioning, left to run wild and unmanaged.

Because the horse is not a ruminant, it must eat high-quality forage to thrive.

The Get Real Program doesn't pull punches. Many will be offended at the suggestion that wild horses need to be removed from federal land. Yet the removal of wild horses and expanded cattle grazing using holistic principles are essential to revitalizing the American West. The fact is that the feral horse is an invasive species. Its population has been left unchecked by man or any significant predators and has grown to destructive levels. It destroys its habitat to the point that no numerous animals can live there.

It is only a matter of time before the horses starve to death on desertified ground. On federal and private land in the west, the destructive wild horse must be substituted for limited duration, high-impact cattle grazing. In the west, the bulk of the land is owned by the federal government and managed (or mismanaged) by the Bureau of Land Management.

Now, cows are a different story when it comes to grazing…

THE COMPLEX CONNECTION FROM GRASS TO COW... AND ITS CRITICAL ROLE IN THE CARBON CYCLE

In a world where cows are one of the environmentalists' biggest targets, arguing that cows can help save the planet is seen as radical. Much is made of the fact that cows emit methane as a byproduct of their complex digestive system, and traditional grazing practices can degrade pastures. This is leading to soil carbon loss and reduced productive plant activity. Because of issues like these, ardent climate activists have made calls for a vegan world. However, the ideas of Allan Savory have totally transformed thinking about grazing, offering practices that can enrich land. There are many YouTube videos (search for Allan Savory) that you can watch to learn more of Mr. Savory's genius ideas on reinvigorating our lands.[12]

Savory proposes grazing methods that can actually reduce the overall carbon footprint of a plot of land by increasing carbon sequestration and hosting cattle while also benefiting the land itself. This effort is consistent with the goals of The Get Real Program to advance the health of our Earth on multiple fronts at the same time. It is realistic and affordable, when looking at the big picture.

The book *Cows Save the Planet,* by journalist Judith D. Schwartz, is one of our favorites on the topic of holistic grazing, so it is a must-read for those truly interested in saving our planet. It offers an in-depth discussion of the benefits of regenerative grazing and provides multiple case studies to back it all up. Despite the potential major benefits of regenerative grazing, the majority of ranchers continue to utilize destructive grazing practices. It is our purpose to

---

[12] Multiple references were from the Allan Savory YouTube TED video: ***How to green the world's deserts and reverse climate change | Allan Savory***. https://www.youtube.com/watch?v=vpTHi7O66pI

"ring the cow bells" to get the attention focused on real solutions like Savory and Schwartz are proposing.[13]

Activists push the idea that cows are significant contributors to global warming and meat eating should be stopped. This could create more injury to our planet than most who don't farm could imagine. It may seem like a simple solution when words are twisted or misrepresented altogether on the impact that our cows have on our precious Earth. However, as an experienced farmer, the founder sees the cow as a tool in the treasure chest for combating the issues referred to as "climate change" and, more importantly, the literal health and sustainability of our planet. Cows are key to the efforts to reverse depleted nutrition in our soils, as well as our ultimate food supply. As any vegetarian knows, it is very difficult to get all the amino acids needed for healthy life from plants, as so many plants or seeds don't have them all.

There is much complexity to the cow, soil, grass, and carbon connection within our environment that is not commonly known. For instance, what do you know about how a cow's urine interacts with the soil? In healthy soil, cow urine soaks in and is used by soil organisms to create humus and other nitrogen-rich compounds. In unhealthy soil, it pools on top and evaporates. Many processes within a cow tie to another, which ultimately leads to the health of our lands, environment, and food supply. All of this is done without anyone noticing or thinking about it, except for a farmer, of course. It is a miracle on top of more miracles to watch the interactions within the environment and farming, involving both plants and animals.

All of the intricate details and interconnections of our environment have been the study of the Founder's life, so that is why The Get Real Program advances multiple ideas on many fronts all at

[13] Schwartz, Judith D., ***Cows Save the Planet: And Other Improbable Ways of Restoring Soil to Heal the Earth***, Hartford, VT, Chelsea Green Publishing, 2013.

once. It will take many actions happening in parallel to create real change, as it is clear that when you change one thing, then another thing will change in response. As we nourish one area, there will be ripple effects needing management in another area. A farmer/rancher has a unique advantage over a scientist in being able to anticipate and understand what is next after the changes proposed here and how the Earth and environment will respond. Cows are the answer to so many areas for multiple reasons, but there is more.

The Cow and Nutrition Connection...

A cow is one of the most beneficial components of our nutrition and of the carbon cycle. How, you ask? Cows are walking hosts for a wide variety of organisms that can create complex proteins from simple sources of carbon and nitrogen. As long as the cow has carbon and nitrogen sources, it can create proteins in its stomach. This leads to greater nutrients ultimately being fed into our soils and impacting our grasslands. Due to the incredible variety of bacteria in a cow's gut, the resulting cow manure is rich in nutrients in a bio-available form (this means in a form available for direct use by plants and microbes). This is why cow manure is perhaps the most ideal fertilizer. The healthier the soil from the fertilizer, the healthier our environment. Dr. Allen Williams has some enlightening talks on YouTube about this. Dr. Allen Williams is a 6th generation family farmer and founding partner of Grass Fed Insights, LLC, Understanding Ag, LLC and the Soil Health Academy.

Vegans like to think that their habits are saving the world, but vegetable production usually involves clean tillage, which generally causes the soil to lose carbon. The fact is that the bulk of land is not suited for farming and is only good for grazing at this time. Vegetables are generally low in protein or have incomplete amino acids. High-quality pasture supporting livestock can make more quality protein per acre than a vegetable crop, while sequestering carbon as well.

According to Savory, the key to fighting CO_2 levels is holistic grazing and grass management where forage-raised animals are primary food sources. Globally levels of beef consumption need to be several times higher than they currently are to combat climate change and rebuild lost land. Savory has said that our problem is that too few cattle are managed correctly. While effective in results, the efforts within The Program on grasslands are complex. For instance, to really convert substantial ground to native grass so that it sequesters many tons of carbon per year will require many cattle to be moved from arid western regions, where feedlots currently are located, back to the Midwest.

It may also seem counterintuitive to continue feeding grain to cattle, but if the feeding is done in a different way, it may be beneficial. By feeding groups of cattle in changing/rotating areas so that manure is incorporated into the desert soil, desertification can be reduced and eventually reversed! Of course, there is more that can be done with utilizing the best farming practices to maximize the use of water and natural fertilizers.

Soil that has been modified with small holes to catch the manure and water will provide a seed bed for grasses and other plants. The holes also serve to catch rainfall, which would otherwise run off easily. We see other solutions too. It is also possible to compost feedlot manure and transport it to areas to be reclaimed. Manure can be put into the catch sumps that capture excess water and debris to support new plant growth along with biochar, which we talk about more in later chapters.

Oh, how we love to talk cows and the impact on the environment. Their bodily functions matter and support both our health and that of the environment. When people drive through the country and see the cows, there are so many unseen processes going on at that very moment that are critical to so much more than the steak dinners and hamburger they produce. Activists try to paint them in a bad light, but we are here to correct that thinking and challenge the lies. When you follow the money, you will see that activists are misrepresenting the effects of cows on our environment and that cows can be part of the overall solution for the rising carbon levels. Just ask your local farmer to explain or, better yet, please visit our website at www.getrealalliance.org and others we reference throughout this book.

Cattle's Natural Challenges Creating Natural Solutions

There are very natural, affordable solutions to our dry, desolate, and desertified areas. In addition to the ideas described in this book there are important innovations being developed in water purification and desalination. Also, in some cases where we see a problem, we already have some answers from experience, not from a lab test that may or may not be able to be repeated across different areas of the world. We, a Farmer/rancher and engineers, have a lifetime of expertise and experience that we are sharing here with you, so that you too can take action and invest in our environment in a way that actually matters and creates real impact.

Cows can be used to solve some of the environmental challenges of today, but not if we start restricting meat. If the government controls our consumption and the breeding of cattle or the economics of it all, our natural resource to fix the environment will be depleted along with our soils and the nutrition across the globe. For example, cattle hooves have a unique pointed shape that allows them to penetrate into the ground as they walk. They leave very small impressions in their wake, which catch rain and open pathways through the trampled crust. In healthy friable soil, such as native prairie, cattle hooves sink deep into the topsoil, mixing vegetation left behind from grazing into the soil.

One of the best ways to heal badly eroded or depleted land is to feed hay to cattle on the land. This strategy was tried on the Founder's ranch, which had some very badly eroded places with no topsoil and was bare of vegetation. In just a few years, the leftover hay and manure had combined to develop grass and plants in those bare spots. As mentioned above, manure can do amazing things as a fertilizer.

There is also an interconnected web of plant/soil life below ground, and above ground plants provide all the energy for grazing animals who keep the grass healthy while grazing as herd animals. Herd animals pursued by pack predators are forced to move often, avoiding overgrazing by partially grazing a piece of land intensely before moving on, leaving the soil and forage with substantial opportunity to rest and regenerate. Even our best golf courses require a little breather time periodically for the greens and require protection from overuse.

The density of predator-pursued herds results in vital ground disturbance, which invigorates new growth. There are areas of the world that still have this type of herd grazing. One is in Alaska where hundreds of thousands of caribou migrate hundreds of miles across Alaska every year.

Let's start to explore more on the carbon-cow connection by talking about methane, which has consistently been part of the false messaging on cows.

The Methane Connection

Much is made of cattle burps! Most of the methane generated by a cow is released in burps from its digestive system through its mouth and nose. The methane is produced by anaerobic bacteria in bovine stomachs that break down indigestible fiber into digestible food. Bovine-based bacteria can literally make protein from the building blocks of carbon and nitrogen.

The bacteria can produce methane, which is a greenhouse gas that persists in the environment for some time. Global levels of methane may be more constant than people think though, because it is all part of a larger cycle where methanotrophic bacteria are consuming methane at the same time in many areas as methane is being released. Note that methane is heavier than air and stays close to the ground when not windy. Also, methane in the air does break down to CO2, which is a less concerning greenhouse gas, trapping much less heat than the methane molecule. It is almost magical to watch nature in play and in the ebbs and flows of its incredible response to change across the continuum. So, while methane has been the focus of the discussion on this topic, we see that the activists don't also share that there is more to the complexity of cows' impact on the environment than just the greenhouse gases that are released.

It is rarely mentioned that these bacteria that produce methane are all part of a larger cycle that includes methane-eating bacteria. There are sometimes simple solutions to complex problems. The Get Real Program advocates to breed and spread more methanotrophic bacteria to consume the increased methane in areas such as the thawing tundra and wetland areas, so that the conversion of methane into CO2 and other compounds is increased. Do you see? Simple solutions exist and we are here to share them!

Scientific law tells us that energy can be neither created nor destroyed, but it can be transferred or changed into another form. It is critical for every cycle to be balanced. It is important to note that cows can't create more energy (i.e., methane produced from digestive bacteria) than what is already available in the field as grass. If left to grow, die, and decay naturally in the field, much of the same grass will also be converted to methane. Cattle are basically transient energy storage vessels that consume and release the same amount of energy that would be released (given enough time) by naturally decaying biomatter.

Methanotrophs

Methanotrophs, methane-eating bacteria, if present in the pasture, are nicely positioned to gobble up the methane as the cattle graze. The bacteria turn the methane into soil carbons and convert some of it to CO_2. None of the aforementioned activists seem to be aware of methanotrophic bacteria, which play a role in digesting methane from any source out of the air at the level of soil and water surfaces.

Methanotrophs can be bred with improved varieties and the land could be seeded with them as a way of drastically reducing methane levels.

These atmospheric methane-consuming bacteria are one reason atmospheric concentrations remain low despite constant emission of methane into the air. As methane is consumed by methanotrophic bacteria, it is removed from the air over time to create a near-balance of methane levels in the air. Populations of bacteria rise and fall based on food supply. Bacteria populations are regulated by an expanding food source when there is growing methane, and once changed into CO2 and there is a smaller amount of methane, the population of such bacteria would decrease. On a macro level the methanotrophic bacterium converts methane to CO_2, which is also produced if the grass dies naturally and decays by oxidizing in the field or is eliminated by use as a fuel source.

Nature never fails to amaze, but this case is especially interesting: a ruminant's breath serves as food for bacteria to recycle the energy in methane. There is a lot of information available about methanotrophs, and it is generally accepted that methanotrophs are especially common in or near environments where methane is produced and that some methanotrophs can oxidize atmospheric methane.

From Wikipedia (Methanotroph): "Their habitats include wetlands, soils, marshes, rice paddies, landfills, aquatic systems (lakes, oceans, streams), and more. They are of special interest to researchers studying global warming, as they play a significant role in the global methane budget, by reducing the amount of methane remaining in the atmosphere".[14]

[14] Wikipedia contributors. ***Methanotroph***. Wikipedia, The Free Encyclopedia. October 21, 2021, 00:23 UTC. Available at: https://en.wikipedia.org/w/index.php?title=Methanotroph&oldid=1050982689

Thus, there may be a case for culturing high functioning bacteria and spraying them onto pastureland to boost methane digestion from cows and other sources.

Let's Love Manure

Cow manure is part of the natural process of regenerating life back into the soil, grass, and then ultimately an enriched food supply. Manure is a wonderful thing in the right quantity, but it is a poison and pollutant when placed in excess. Ideally, manure needs to go back onto the land where the cattle feed was raised. Unfortunately, in our system, corn is grown many miles from where cattle are finished on grain in concentrated lots. So much manure is produced that it is not economically feasible to haul it out to the land that needs it.

Another issue with redistributing the manure is that much of the nitrogen in the manure oxidizes into the air or leaches away. Currently, there is no reward for being a good steward of manure. A relentless drive to the lowest cost of cattle finishing results in bigger and bigger feed yards. Larger yards minimize management costs per head and allow use of the biggest equipment. Though they have some economic merits, confined animal feeding operations catch a lot of bad press from animal rights and environmental groups for many valid reasons.

The Cow Market and How it Ties to Carbon Solutions

There is a wide range of prices paid for cattle, and visible quality does raise the price; however, with few buyers and lots of sellers, prices are rarely very good. Auctions in which none of the sellers will take their animal home and there are only a few buyers are hardly fair to the sellers. The Founder lost money in the cattle business until he got away from the sale barn and started selling to a local end-buyer (i.e., a small slaughterhouse and meat store).

The commodity exchanges where prices are set for many things are even worse than the auction. This is because fictitious supply can be created by putting down a very small cash deposit to sell something that you don't have to buy in the actual market. On the commodity exchanges, the product trades on the exchange in high volumes compared to actual supply. There are a lot of producers and speculators, but only a few real buyers that make money keeping prices paid in balance.

Buyers set the rules, especially when there is plenty of supply and no dominant producer. The system is clearly broken when producers only make money if they are a below-average-cost producer. The average cost producer has to have another source of income to live on. This is one reason the commodity beef business produces low quality beef, giving beef a bad name.

A grass-fed system where quality meat producers would have a stake in the business of selling beef to stores or customers directly would be much better on all levels. There is a problem though. To finish all the cattle that are currently in feedlots will take a lot of high-quality land that can grow good forage to produce top-quality grass-fed beef. The prairie grass feeding system detailed later in this chapter can provide America and the world with top-quality beef in a way that both sequesters carbon and produces sustainable energy.

When one flies over the American Midwest, they see a grid of squares with green circles in them. These are square-mile or one-quarter-square-mile tracts with center-pivot irrigation on them. Due to the arid climate, growth is minimal on the unirrigated corners of these fields. Imagine the pivot irrigation unit that makes the circular pattern, being replaced with a unit that also harvests a portion of the field of native prairie grass so that it can be fed to cattle housed in the unirrigated corners of the field. It is desirable to harvest when the sun is shining as the feed is of best quality at this time, so it is logical to power such a system with the sun. If grain crops are going to be replaced with native prairie grass and holistic and regenerative grazing is the new norm, we will have to move forward with new thinking and not remain stuck in the old ways and with limited thinking.

HOLISTIC AND REGENERATIVE GRAZING OF GRASSLANDS

Grasslands cover huge amounts of the globe and people often underestimate how much carbon grassland can sequester if managed properly. Grassland that is not grazed at all declines just as badly as grassland that is improperly grazed using traditional methods. The key to healthy grassland is to simulate the activity of herd ruminants under pack predator pressure, which is how grassland evolved. This is what Allan Savory has promoted successfully on millions of acres around the world.

Savory has restored degraded land to lush, carbon sequestering productivity. Ironically, his methods also produce much more beef per acre, so universal adoption would help solve the carbon crisis *and* increase the food supply. The misguided anti-meat crowd really misses the reality that the largest amount of land is grassland, and most of it is mismanaged either by continuous grazing or lack of grazing. Getting rid of cows would make things worse. We need to focus on grass finishing them, which means getting the cows ready for their ultimate slaughter. The best use of our grassland is engaging in holistic grazing by keeping the cattle out in the pasture and moving them periodically.

As mentioned, there are calls for a sharp reduction in meat eating because cattle are a blight on the environment and contribute to climate change. But the real problem lies in incorrect grazing by too few animals. Holistic grazing, as advocated by Allan Savory, is proven to remove CO_2 from the air and build vital soil carbon and moisture. The key is to keep the grass healthy and to keep the soil covered with vegetation and mulched plants so that it retains moisture and gives the grass room to regrow. The Savory Institute has powerful documentaries on the internet that demonstrate how much progress can be made and how critical it is that we change our ways.

The Holistic Grazing Solution

Holistic grazing ideally keeps the cattle bunched together and uses the loafing time to concentrate manure and urine in a small area. The result is high fertility areas. Cattle naturally congregate together for protection against predators, even though in most cases modern cattle face no predators. In many western areas, keeping cattle overnight on flat ground along a dry creek bed can restore the formerly green meadow to life. Ultimately, these grazing practices can create soil that soaks up rainfall and make creeks flow again. Otherwise, creeks will remain drainage ditches, carrying a flash flood of rainfall then quickly drying up again. There is vivid evidence of this in videos by Allan Savory on YouTube, including *"Regreening the Planet..."* and *"How to Green the World's Deserts..."*.

In highly fertile areas with good rainfall or irrigation, it is hard to graze cattle without substantial losses due to trampling. Trampling is good, but only when the grass is sparse and some ground is uncovered. Cattle tend to avoid stepping on grass if it is not solid, but they have no choice when the grass is continuous and there is no bare ground.

There are currently hundreds of millions of acres that could be growing grass but are instead becoming desert, sometimes gradually and sometimes quickly. Most of the desertifying land is owned by the federal government and could be reclaimed into carbon sequestering grassland. The major effort of digging small water catchment basins across the land, along with introducing organic matter to provide a seed bed for grasses, is necessary to reverse desertification. This is what it will take, but we have to have willing leadership to make the changes necessary for the use of the federal land and investment to get us there.

Another major effort is needed to significantly lower CO2 levels in the air. We can't have large amounts of land sequestering no carbon and not storing any water in the soil. Land in the western United States is so degraded that it will take a big effort to restore it to vitality. We could go on and on with the use of our grasslands as part of the solution for multiple environmental challenges, as well as many other benefits to America.

Investment Is Required

When we think of reversing hundreds to thousands of years of decline, we must be prepared to invest time and money in the project. There are millions of acres of land that have been regenerated very economically through holistic management when there was some grass left. However, in some cases, especially where the wild horse has denuded the ground, more extreme measures will be needed. We advocate for grass-fed beef raised in a holistic manner.

The vast majority of grassland needs to be regeneratively or holistically grazed in order to have the impact that is needed to manage our carbon levels. This will increase production dramatically but also require much more management of our natural resources, which will take investment and smart farming practices. There are so many solutions that farmers do and do not control.

There is something wonderful about raising and managing a herd of cattle. Ranchers love to do it even though most currently lose money using traditional methods. There is a lot of pastureland that sits empty except for wildlife. The grasses are degrading, or in many cases disappearing. With proper grazing and better management, this land can be productive.

To do all of this, the pricing of cattle needs to change. Successful grass-fed beef producers know they have to avoid the commodity beef channel, which is generally unprofitable for producers. Cattle are being sold today for the same price as 50 years ago, while equipment and other costs have gone up dramatically.

While people can fantasize that it's ideal for a cow to be outside all the time, a cow is miserable being outside when it is too hot, wet, or cold. This is particularly true in the fertile American Upper Midwest, where winters are very hard on an outside animal. The Get Real Program advocates for a change to grass-fed beef inside feeding barns in these areas; with native prairie grass replacing some of the corn acreage.

The Founder knows the business of farming and what needs to be done, and shares more throughout this book and on his website www.fullofideas.com. While holistic grazing is critical to our success, there are other amazing solutions that we have a vision to implement. Right now, our grasslands are not being optimized for all that they can offer, as there are many needed enhancements, even without remineralization efforts in our soil. We are excited to share more below, as prairie grass management will complement the holistic grazing efforts.

HIGH-INTENSITY PRAIRIE GRASS MANAGEMENT

There is so much land available to be put to its greatest use, rather than letting the resource waste away without being utilized and becoming sterile from lack of management. The Get Real Program knows how to manage the land in a way that creates increase. Beyond the expertise of The Program, we share so many more resources to learn about what is happening around you and ways to do something before the grass is gone.

The solutions brought forward by The Program can prevent soil erosion and increase carbon sequestration while minimizing water loss and moving America to a more sustainable grass system. There is so much more to this area of study than we can do justice to within the bounds of this book, so please make sure you review all of the materials provided for added reference throughout this chapter and the Founder's website at www.getrealalliance.org. As America moves to greater prairie grass management, other benefits are part of the package, for instance, having small slaughterhouses located where there are clusters of cattle being grazed on native prairie grass so the cattle could walk to the slaughterhouse for their final day, instead of having a traumatic experience packed in trucks and more.

Researchers generally study things that are currently achievable so their research can be adopted. Commercial agricultural equipment for harvesting native prairie grass at a high height is currently not available, so most research that has been done has used available equipment that cuts the plant at a low height. Cutting native grass too low to the ground removes most of the leaves and stops photosynthesis. This sharply reduces plant productivity, but even so, the forage production is high and of good quality.

The prairie grass systems might amount to one percent of the total grazing ground in the U.S. but could produce a great deal of grass-fed beef and energy. You can go to www.prairiegrasssystems.com for more details on the prairie grass system and its high food and energy production potential. This is an area needing great investment and study to challenges the current practices and threats to our food supply at a high cost.

FOI Group LLC, an invention company founded by the author, has a way to convert some of this highly productive midwestern land that is currently growing grain crops back to its original highly productive native tall prairie grass polyculture. The solution is a rotating system that harvests prairie grass at an optimal high height so that growth is maintained in a sequential way that allows the grass to recover for a month or more before it is cut again.

Cutting the prairie grass at half the plants' height leaves lots of leaves to keep feeding the soil life, and the plant regrows quickly as the roots are not depleted. Unlike traditional haymaking, where a tractor and equipment compact a lot of soil, the circular harvesting system can have widely spaced tires traveling in set arcs to support the structure so only a small fraction of the ground is driven on and compacted. This system could more than double forage production compared to traditional methods.

Lots of forage is left in the field to keep the prairie grass healthy during the growing season. It should be harvested after a frost to prepare the plant for the spring. Prairie grass needs to be left at a fairly high height during the growing season; however, the grass regrows from its root base in the spring, so it is important to not leave dead forage, which blocks sunlight from the new growth.

The harvested forage after frost can serve as part of the forage needed for winter. Prairie grass is a bunch grass with a large root mass and tops that splay out to the side to cover the ground, even if the plants are spaced apart. Thus, after frost when the grass is cut low, winter forage can be planted around the plants for winter feed in areas where it grows.

The native prairie was a super-efficient mixture of plants – not a single species as man often tries to create. Some estimates are that there were hundreds of different plants working together to create more growth above and below ground than a single species could. Tests have shown sharply higher total forage production out of polycultures with production rising as many species are added.

The Prairie Grass Solution

There are many advantages to harvesting grass mechanically and bringing it to confined cattle to eat instead of serving a grain diet as is currently done to finish cattle – meaning they are fattened up and ready for slaughter. Prairie grass mixtures are highly palatable food for cattle but die out under continuous grazing as was done by early settlers. Grass-fed cattle are really better food nutrition for people, with more desirable fats; if properly fed, there is adequate fat yielding very tasty beef. With grass feeding it takes longer for an animal to get to optimal weight as a mature animal that then puts on fat. Many small grass-fed beef producers rush their animals to market and often don't have truly top-quality forage year-round to produce optimal beef. The Get Real Program has a comprehensive plan and innovative solutions through inventions to improve the farmers' ability to advance towards greater use of prairie grass as a tool that enhances our food supply.

If a tall prairie grass system is going to make more money than a corn-based finishing system, the cattle cannot trample the grass and contaminate their feed with waste. Flies and parasitic worms that grow and spread on manure cost cattlemen a lot of money. On low-density pasture that is regeneratively or rotationally grazed, these issues are alleviated.

The prairie grass solution (and global opportunity) is an opportunity that we believe in and we also believe it can be properly done with farmers' and ranchers' hands organizing the effort. As a beginning, we are supporting a program to develop a prototype of the prairie grass harvesting system that will show how beneficial it is and how it can be part of the solution to save our land. As engineers and a farmer/rancher, it is exciting to drive for solutions for change that will impact and save generations to come... save them from disease and protect their food supply, while also sequestering CO2.

The Challenges of Excess

While many tracts of land are well suited for regenerative grazing, there is highly productive land where grazing cattle leads to waste of forage and possible food production due to trampling and manure contamination. Trampling and manure help build up land that is in bad condition but are wasteful on highly productive land. This is especially true of land that is very poorly drained.

Highly productive land that produces enough forage to support several animals per acre suffers significant losses due to the sheer numbers of holistically grazed animals trampling much forage and overly contaminating significant areas with manure and urine. In many cases, the most productive soils are high in clay, which holds nutrients well but makes the ground remain muddy for days after a rain. Many tracts and soils are fairly flat—holding water too well—

and farmers have had to install underground drainage systems to dry the ground enough to be able to productively farm there. However, cattle can do a lot of damage when grazed intensively on muddy ground, compacting the soil and sharply reducing productivity.

In the Midwest, land that is highly productive often requires underground drainage to allow it to be farmed efficiently. Lots of chemical fertilizer leaches out through these drainage pipes. American agricultural land tracts range from land that takes more than one hundred acres to support a cow to tracts that can feed several per acre. The Founder's ranch had a wide range of soil types, from low carrying capacity to very high carry areas. On very fertile land, a massive amount of beef can be produced from an acre, but proper management must be exercised to prevent waste due to damaged forage and excess manure.

Advancing More Pasture Solutions

FOI Group, LLC, has patented a new system to selectively harvest tall prairie grass in a way that keeps it tall and growing and doesn't trample it with cattle or regular equipment. In the prairie grass-harvesting system, a lot of the forage needs to be dried and stored for days when the forage quality is low, like on a rainy day or for the winter season when no grass grows. The innovative use of a solar thermal process to make some of the power needed to run the system, with the additional use of the waste heat to dry some of the forage for later use, creates top-quality forage all the time. Waste heat from a biogas-fueled power system could also be used.

It may seem strange, but there are times when the forage is too wet for optimal digestion. If feed is too dry, a cow can always add water to her stomach by drinking, but if the feed is too wet, the cow is helpless. Remember that a cow regurgitates and chews over the daily food to fully digest it. Cows spend many hours either in the shade on a hot day or the sun on a cold day chewing their cud. If the feed is too wet, the digestive system doesn't work well and some nutrition escapes in a liquid and smelly manure. That is wasteful.

Given a choice, a cow grazing on very wet forage will seek dry grass to mix in for a more balanced moisture level. Having a dehydration step as part of the prairie grass system produces top-quality feed that can be saved for times when the system doesn't run, such as on cloudy days or in winter. Forage that has been dried artificially retains more of its vitamins and nutrients that would be leached out by traditional sun drying.

As part of The Get Real Program solutions, we have been working on a system to harvest high growth, tall prairie grass for use in feeding cattle. This tall native grass is fed to cattle in a confinement area for rapid weight gain when finishing, thus producing optimal meat. Confinement feeding also allows the manure to be gathered and digested for biogas before being returned to the land by soil injection.

More on Waste and Famine

It is imperative to create more opportunities to use our natural resources to their best use and outputs. If we do not, we are on the road to further the sterilization of our soils and ultimately to create food supply shortages. FOI Group's high-height harvesting system should offer significantly higher production of food and fuel per acre, as well as keep the plants continuously feeding on soil microorganisms without stopping growth and photosynthesis, thus sequestering more carbon into the soil. Going from row crops where the ground has nothing growing in between to having plants feeding the soil for all of the growing seasons will be a game-changer for carbon sequestration. Turning sunlight into money is the goal of farming and using more sunlight can mean more food and fuel, thus more money.

THE NEXT PHASE... BIOGAS

By eliminating wasteful practices involving our natural resources, we can advance our energy independence. Yes, again, cows are part of the solution. We will discuss this concept more in later chapters, but wanted to introduce it as part of the comprehensive solution to the management of grasslands.

Basically, a cow eats forage and turns a lot of it into meat. Along the way, some methane is emitted, and its manure can be digested in a biogas digester to create usable methane to provide renewable energy. The world is currently exploiting geologic natural gas at an extremely low price that often barely recovers the cost of drilling a well. We give no value to the cost of making it in the first place and no consideration to what we will do when it is gone.

We have been very wasteful in general, letting the bulk of nutrients and value in animal manure oxidize and leach away, leaving the ground littered with dried-out manure piles. We need to globally practice biodigestion followed by composting to make good use of the manure that comprises much of the volume of forage that cattle consume. In the U.S., rebuilding the desert Southwest is a massive undertaking, but it is one that needs to take place to lower CO_2 levels and increase the productivity of the Earth. Sadly, much land is so badly eroded into deep gullies and other high slope areas that reclamation is impossible. But there are lots of fairly flat lands that can be reclaimed, and there, biogas production from manure in some areas can be part of the solution.

Regulators have provided for utilities to pay a big premium for renewable biogas over the price of geologic natural gas, which makes biodigestion economic. Growing lots of tall native prairie grass may seem like a drop in production compared to corn, but with an integrated system, it is actually far more productive and profitable. The proposed carbon sequestration payments discussed in a later chapter will make native grass more profitable overall as the carbon sequestration will be substantial.

Substantial beef can be produced per acre, nearly equaling or exceeding the weight that can be produced by feeding on corn, and the collected manure makes a lot of biogas that has good value. On top of this, there is the ability to incorporate the biodigester effluent back into the land for fertility, dramatically cutting the need for fertilizer. Native grass needs no herbicide, and with the addition of winter crops, the soil is fed year-round, keeping soil life at work building humus and other wonderful carbon-based compounds.

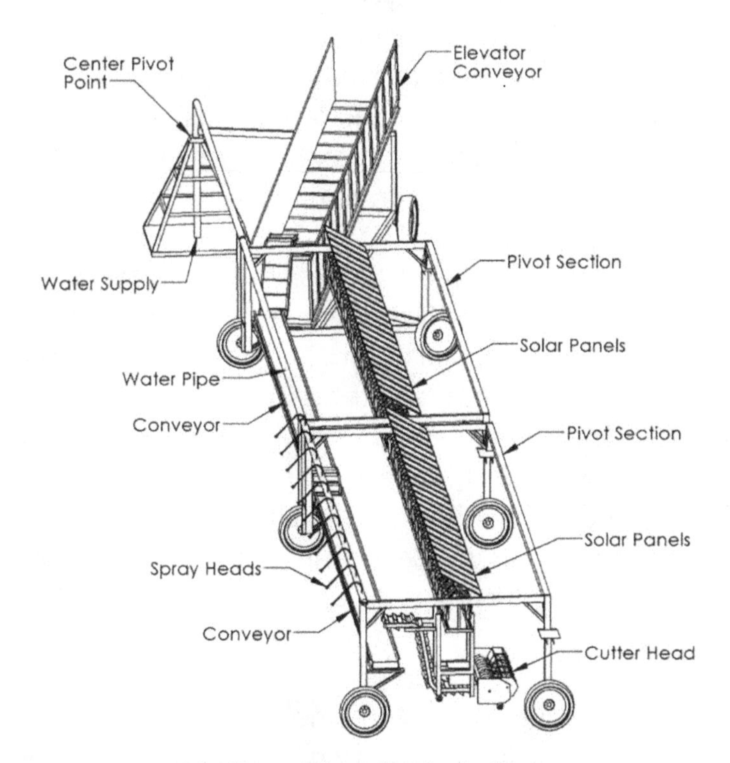

Irrigation and Harvesting Center Pivot

This figure is an illistration of a center pivot with two harvesting sections, more sections can be added to fit field and crop requirments.

Figure 2

VULNERABILITY IN OUR FOOD SUPPLY

This book proposes exactly the opposite of what many concerned climate activists propose; it argues for far more cattle raised in a holistic way that will result in sequestering carbon and making the world a better place. While it is ideal for food to be produced where it is needed, the fact is that many areas of the world have lots of people but lack good soil or a climate conducive to raising much food. Holistic and regenerative agriculture training worldwide can improve this situation.

Realistically, America needs to export meat to the world instead of grain. In terms of nutrient recycling on farmland, it is far better to export meat than corn. Animal waste, or manure, contains soil rebuilding nutrients that end up in another country when grain is exported to feed their livestock. Exporting grain strips nutrients from land at a much higher rate than exporting meat, and it doesn't allow for animal manure to return to the land that grew the crop or grass.

Right now, the world lives with almost no stored food, and a major disruption would lead to starvation. Raw meat is a short-lived, perishable commodity that requires refrigeration or freezing to store. Neither storage method is optimal in the long term without spoilage or quality loss. With increased beef production, some of it could be freeze-dried, made into jerky, cured, or canned for both export and preparedness in case of supply disruption. We may face times when food production is disrupted, such as by shutdowns due to COVID-19.

The world currently lives in a "just-in-time" mode, with no stockpiles of food in the event of disruptions. If we are going to move towards a more equitable and prosperous world, increased meat production is a step in the right direction. Meat is something that people crave as they gain more affluence; however, meat must be provided in a sustainable way. It can't be sustained in the way food is currently raised with grain being grown and shipped many miles to a distant point where it is fed to animals, with their manure often being wasted. As a nation, it is even worse when the grain is shipped overseas, exporting our land's nutrients, leaving behind bare, stripped soil to be eroded into the ocean.

Artisanal butchery can determine whether a carcass is top-quality meat or better suited for chopped or ground beef. In the age of the internet, we really need full tracking of meat so that a consumer can register a complaint, and the rancher who raised the calf can be informed. Currently, animals change hands at auction with no one knowing where they came from, and animals from many operations are combined into truckloads of cattle going to the next stage of cattle raising.

Current thinking on climate is that the Earth is going to warm, and crops are going to continue to grow at higher latitudes in North America. However, last year, the Upper Midwest crop planting was delayed due to cold, wet weather, and harvesting the crops was delayed or not done at all because of wet conditions due to an early fall. In 2020, there was a late freeze and snow in April that also damaged the northern crops. Time will tell if this continues to happen, but we could see the Upper Midwest become less hospitable for growing corn due to a shortened growing season.

As farmers, we are sounding the alarm for anyone and everyone to become concerned about the current state of our food supply, which is *your* food supply. Everyone is impacted, not just the meat eaters or those living in urban areas. There is an interconnectedness of everything and everyone that cannot be undone when we look at our environment and our own nutrition. It is all connected, so what is challenging and harmful to one will also be challenging and harmful to another.

No one can escape our interconnectedness. We have one environment to share globally, which impacts our food supply, the soils for growing crops, and the air we breathe. It will also impact future generations in a very dire way if we don't act now. Are you convinced you too can have an impact by your action or inaction?

INNOVATIVE, YET DOABLE SOLUTIONS

Properly managed cattle are essential to reducing CO2 levels and preventing the growth of deserts. Go ahead and eat beef AND buy grass-fed beef when it's available to encourage more growth. We'll be in a very bad place if we don't manage the grasslands in North America for productivity and cattle are an essential part of grassland preservation. It is imperative that we encourage holistic grazing in other countries too, especially some of those in Africa and regions that oftentimes have degrading grasslands due to poor management practices. Building a prairie grass harvesting system will be a game-changer for our environment by increasing productivity from the land, and will generate more cash flow for the landowner and for the prairie grass system.

The Get Real Program's vision is to see prairie grass harvesting systems, and clusters of them, around small slaughterhouses where the cattle have a much less traumatic end-of-life compared to current practices, and thus greater quality meat. The vision includes better employment experiences for the workers in the slaughterhouses to help grow our rural communities. Also, The Program envisions selling the meat for a better price as branded products, instead of the discounted commodity market product that's in the grocery stores now. The cow and meat pricing structure merits another book of its own, as it is now very "backwards" in its rewards to producers. Get Real Alliance fights for the hardworking farmer and rancher doing the tough and passionate work of getting meat to our table. We believe that people would pay a premium for good food to advance the lives of the hardworking people managing our cattle and our important sustainable grasslands day-in and day-out.

In *Holistic Management*, his book about brittle environment restoration, which is essential for man's survival, Dr. Allan Savory states:

> "That we have been unable to stop the deterioration is not surprising since approximately two thirds of the Earth's landmass is brittle to some degree and our management has not catered to that fact. Not until now have we known what to do. With desertification and climate change now threatening our survival as a species and with the bulk of the world's brittle grasslands and savannas deteriorating so rapidly, we are only left with livestock, managed to mimic the behavior of once vast wild herds, to begin the urgent process of restoration. While only those actually managing land can implement the changes necessary to reverse the desertification process and regenerate carbon storing grassland soils, they will need the support of the rest of us, including policy makers, to do this effectively."

We strongly recommended that readers buy Savory's book and support his organization, which is really making a difference in the world. See their website at https://savory.global/. Savory's work is evidence that one of the solutions to the global climate crisis involves more cows properly managed to mimic the behavior of large, wild herds that move continuously to stay ahead of pack predators. This allows the grass and its roots to regrow and thrive instead of being killed out by continuous grazing, or worse, no grazing at all.

We all need to buy grass-fed beef when possible and push for Savory's principles to be made into federal regulation for the vast lands managed by the federal government. Supporting small meat producers is another great way to encourage better cattle management. We should be exporting beef instead of corn because exporting corn depletes the soil and encourages wasteful carbon emitting practices.

For a long time, Savory's message was proclaimed through the alternative agriculture press and ignored by the conventional chemical agricultural press (i.e., a farming magazine that is often free and filled with lots of chemical ads). But things are changing. In the March 2020 issue of *Progressive Farmer,* there was a nice article about holistic grazing that said:

> "Savory, a native of Zimbabwe who has lived in America for decades, was and is controversial. Savory developed a method for planning livestock grazing that enabled pastoralists to simulate the effects wild herds once produced on the land. But more than grazing, management had to change in order to transform landscapes. The idea of holistic management arose from that insight."

The rancher featured in the article had this to say about progress:

> "As far as rotational grazing goes, NRCS (Natural Resources Conservation Service) and other government agencies have come late to the party. When I started, they weren't very supportive. To their credit, they figured it out, and now we're working together to help everybody understand this."

Much has been written on holistic grazing and its potential to reduce CO_2 and improve soil health. The efforts involving prairie grass are also justified through real-life research. If you harbor any doubts, there are many books and publications available to support these claims. The Acres USA bookstore is a great resource. Readers are invited to search and find more proof about these methods.

The evidence is clear. We can dramatically increase carbon sequestration with proven practices. While these practices are known by some progressive farmers, they are largely not adopted on most grassland. Challenges exist, but with a little ingenuity from farmers and engineers, nothing is unsolvable. We need a different way to use these high-production fields that does many things: produces a lot of quality food, builds soil carbon for carbon sequestration, makes biofuel and energy, recycles nutrients, and provides profitable year-round employment.

We can see results from a small investment leading to the large impact of returning to a healthy, invigorated environment and food supply. There are regenerative grazing operations that show a one-percent increase in organic matter per year over years. There are 10 tons of more carbon per acre, which amounts to 30 tons of sequestered CO2. The American Midwest could produce well over 1,000 pounds of beef per acre per year if converted back to well-managed, tall native prairie grass!

This can be achieved through regenerative farming and the application of compost and biochar. This land could sequester huge amounts of CO_2. The CO_2 would remain in the soil as the soil would never be bare during growing periods, unlike a corn or soybean crop cycle. Sequestering 10 tons of carbon at a payment of $50 per ton is $500 an acre a year of revenue for doing things that make the land more productive.

Using the proposed methods, ranchers can produce twice as much beef per acre as the area average while sequestering a lot more carbon. We can make a big difference with more food, which is better for you, and also with less CO_2 in the air with holistic management. We can really help feed the world and reduce CO_2 by changing how we manage cattle and grasslands.

Some people mistakenly think that not eating meat is good for the environment, but the opposite is true. It is how the meat is raised that makes all the difference. We actually need twice as many cattle, all being managed holistically, here in America. You can help by supporting Savory's organization, the Savory Institute, and buying grass-fed beef when possible. If you don't have a local butcher shop, you can buy online at https://localyocalfarmtomarket.com/.

Go to getrealalliance.org for more information and to learn how you can support the effort to change our disastrous path to a positive one.

CHAPTER 7

RICHES IN REFORESTATION AND THE MAGNIFICENCE OF OUR EVERYDAY TREES AND DESERT SHRUBS

Our trees are in trouble, which means our very existence is in jeopardy if we do not do something… and fast! There are trees everywhere, including deserts, but it is in the forests where we can do the most good for our planet and the environment. Even so, no tree area can be ignored in our plight to create real changes for sustainability.

The Get Real Program is proposing that we: 1) change the way we harvest timber, especially from federal lands that aren't managed tree farms; 2) adopt The Program's aerial harvesting ideas to minimize the destruction of trees that aren't being harvested; 3) generally keep the forest growing, cleaned up, and sequestering carbon, and 4) remineralize new and old tree areas for increased growth and health. There is a lot we can control in regard to our environment, including changing man's practices of clear-cut logging to avoid the massive loss of topsoil that it causes, as well as other negative outcomes. As with so many other business practices, logging is generally done in the cheapest way that is very destructive to the environment.

Trees and forests are rich in resources that impact many critical areas of our lives, even those in obscure rainforest areas that we don't typically think about or see. It is imperative that trees get the respect they deserve for their part in the carbon cycle, which impacts our every breath. Revitalizing and remineralizing our forests everywhere is key to advancing carbon sequestration efforts.

We must replant areas where the trees are all dying and diminishing, as well as remove deadwood from the forest to reduce the danger of forest fires. As we remove the deadwood, we must repurpose it and fight nature with nature. If we do, America and the rest of the world could see material, positive impact on the rising CO2 levels. Effective management of and planting more of this precious resource can create a significant opportunity for America to advance in cleaning up our environment.

Love for the Trees

Trees are fantastic for the environment, as they alter it in so many positive ways. They provide more than just the pulp and lumber for paper products and our homes and furniture. Even desert-hardy trees have a place in helping the environment, as they sequester carbon and provide habitats.

According to National Geographic: "Rainforests are Earth's oldest living ecosystems, with some surviving in their present form for at least 70 million years. They are incredibly diverse and complex, home to more than half of the world's plant and animal species—even though they cover just 6% of Earth's surface."[15]

Climate Cooling Impact

One can detect how much cooler it is under a tree partly due to the transpired water vapor that cools the air. This effect is known as evaporative cooling. In fact, in tropical forests, the humidity from

[15] https://www.nationalgeographic.org/encyclopedia/rain-forest/

the transpiration of trees can be near 100 percent. This makes the dew very heavy and triggers near-daily afternoon showers. Water vapor acts to moderate the climate as it has very high specific heat. It takes a lot of added heat to raise the temperature of water vapor. It can be relatively comfortable at a very high temperature in the desert as the dry air has low specific heat and can't transfer significant amounts of heat to a person or object compared to humid air.

Climate activists have appropriately changed their terms from "global warming" to "climate change" because the evidence did not back up the former terminology. Average temperature may rise in areas with lower CO_2 levels but higher humidity. The term global warming was really not impactful because temperatures would have to rise dramatically to cause problems like sea level rise (which historically is a naturally occurring effect from glacial and long-term solar cycles).

LOVE FOR THE FORESTS

There are forests of all types in the world. Forests range from the very dense tropical rainforests, where ground is never visible and vegetation is so thick it can't be walked through, to arid high-desert forests where the trees are scattered and small. What they have in common is the amazingly durable tree, capable of existing in many environments, even on almost-solid rock with no real soil.

Our rainforests cover a critical mass of this Earth and need to be protected, just as we protect our endangered animals. Rainforests are so named because a large area covered in a lush canopy of trees creates rain due to the amount of moisture transpiring from the leaves. People often think of rainforests as a tropical phenomenon, but they exist in colder areas as well, such as the Pacific Northwest of the United States.

Man's Effect on Earth's Sustainability and Our Trees

America is currently on an unsustainable path of logging old-growth forests where the trees are 100 years old. We are depleting that resource so much that we may reach a point where we won't have many old large trees in the future. We could have shortages of lumber, which causes other problems, like what we saw earlier in 2021 with lumber prices spiking three times higher than they were last year. While there is currently a large number of trees growing overall, there is mismanagement of the forests with too much density in some areas where small fires have been put out over the years, with no clearing of deadwood or thinning of overgrown trees and clear-cutting in other areas. In many areas, this leads to explosive large fires and in others desolation from clear-cutting, both causing erosion and other problems. Urban sprawl and the tendency to put too much land under pavement is causing a loss of forested areas near cities and suburbs and is in turn causing larger amounts of destructive flash flooding and devastation of mega-floods in some low-lying, flat areas. An example is the Houston Metro area (multiple counties and upstream areas are involved), where such tremendous "pavement propagation" has led to an uncontrollable flooding situation when there is a high-rainfall event over a large area as with hurricanes and tropical storms. Even lightly forested areas preserved in and around the metropolises would retain significant parts of rainfall events and reduce the mega and flash floods that are so costly.

Trees Are Everywhere

Because trees don't produce protein for the bulk of their structure, they require very few soil minerals to grow. A ton of wood burns down to just a small amount of ash, which is comprised of compounds and minerals taken out of the soil. The vast majority of a tree's mass is carbon from the air and water from the environment (note: cellulose and lignin are made of almost entirely carbon, hydrogen, and oxygen atoms).

As we discussed in the last chapter, grasslands are one of the most resilient environments in arid climates. It seems almost at odds to talk about dryland trees instead of dryland grasses. Grass requires more nutrients than trees, though, and this is because grasses produce protein while trees are largely protein-free.

Trees can grow on very infertile soil as long as they have water. For example, the mesquite tree can thrive in a very dry climate. The creosote bush that covers much of the Southwest is another example of a non-grass plant that works well in dry climates.

A great thing about trees is that they can cling to steep places where shallow rooted grass can't. While there are areas in the arid western U.S. that are flat enough for grazing, with the poor environmental conditions, it may be better to plant improved varieties of trees in targeted areas first. If the U.S. is to reduce its use of fossil fuels or be ready for the time when the large reserves are depleted, the country needs to be producing lots of bioenergy from lands that are not used for food.

Our Trees and Forests Are Dying

The forests of the American West are so badly decimated by insects and disease that most of the trees in many areas are dead or dying. Drastic and potentially expensive action is needed to prevent more mega-forest-fires that release lots of CO_2 back into the air. Much of the terrain is so rugged and the roads so few that access on the ground isn't feasible. Thought must be given as to how to enable the wilderness areas to be returned to health. Such a change for the better will help pull down large amounts of atmospheric CO_2 and prevent the emission of CO_2 from excessive decay.

Man hasn't made soil and forest health a true priority. It needs to become one if CO_2 levels are of concern because healthy forests can be a major sink of carbon while unhealthy ones emit significant carbon.

Healthy trees live for a long time. Utilizing innovative solutions while also investing in effective carbon sequestration efforts through funding charitable organizations like www.getrealaboutclimate.org, we can turn the forests into massive carbon sinks and restore the lovely national forests to health.

Planting Is Only One Part of the Overall Solution

Planting trees is a great thing and needs to be encouraged where conditions are right. A great example of this is a small charitable organization, the Archangel Ancient Tree Archive, working with the famous redwood trees on the Pacific coast. The organization's mission is to propagate growth trees, reforest, and archive ancient trees' genetics (see https://www.ancienttreearchive.org/ for more information and to support the organization). They are cloning the redwoods and replanting them in their natural range. Redwoods are a long-lived and fast-growing tree that can sequester a lot of carbon, especially with rock dust remineralization of the seedlings.

The organization *is* incorporating rock dust in their potting mix to grow strong, healthy seedlings. While small organizations like this are praiseworthy, a global reforestation effort is needed. A global reforestation program could make huge strides to roll back CO_2 levels while improving the world's overall environment. Ideally, this would be done by efficient non-profits instead of inefficient government programs.

Research Is Crucial

Innovation must be prioritized to preserve and expand the world's forests. There is a need for research to develop improved varieties of trees. Potentially dramatic breakthroughs could result, including fuel coming from trees.

With fossil fuel so cheap, there has been little interest in developing plants that produce liquid hydrocarbons. These plant-sourced hydrocarbons could be made into liquid fuel to replace gasoline or diesel. A tree similar to a maple could be developed to produce large quantities of high energy liquid hydrocarbon sap. Other plants could also be researched for such a purpose. It might be a cactus-like plant that produces sap that can be refined into fuel. The main thing currently preventing such innovation is the super-cheap availability of fossil fuel. See the chapter on smart energy options for more discussion.

Knowing what's at stake, consider supporting the organization Remineralize the Earth www.remineralize.org in a big way to help them in their quest to get more research completed. Positive results have been seen when using rock dust at the planting of trees with dramatically better growth, but little recent work has been done on remineralizing existing stands and comparing growth. Research takes money, and much more support is needed to quickly prove effectiveness in more areas of the country and world.

Unfortunately, there are many activists who are really hopeless about the future and paint a dark picture of future life on Earth. Jim Griffin makes this point in an article he wrote for the online publication at https://timberupdate.com. The prices and income data cited in the November 21, 2011 article show timber property's current marginal profitability, which demonstrates the opportunity that could be driven by providing funding for private global efforts for faster forest growth and a better environment in our future.

> "The stand is *growing* at 3.3 *tons per acre per* year (1.25 cords/*acre* volume *growth* potential). The product ratio within the stand is 2/3 pulpwood *trees* and 1/3 chip-n-saw. Assuming $7 *per ton* for pulpwood and $15 for chip-n-saw, the annual value *growth* for the existing 450 *trees per acre* is $31.86/*acre*/year."[16]

[16] https://timberupdate.com/blog/thinning-timber-can-earn-350-per-acre-and-increase-annual-harvest-by-25/

SOLUTIONS AND MORE

Water, Land, and Our Trees

One of the ways arid land has been restored in Africa involves digging small indentions in the ground and putting nutrients or mulch in the hole along with a plant, either grass or tree. The depression catches water and holds it, instead of allowing it to rush away in a flash flood, as usually happens on land that is crusted over. There are those who advocate against any manmade changes to the natural environment and will object to replanting the desert with vegetation, but creating a grid of water catchments across the landscape will alter it in a positive way. Runoff after a rain will be lessened, and more water will soak into the ground, allowing more plant growth. The Get Real Program advocates for a major effort to minimize soil loss and prevent flash flooding, landslides, and damaging erosion. After a year some desert trees can live with watering only once every 3 or 4 weeks. So initial care can be provided for a year followed by only intermittent watering until a tree becomes truly self-sustaining in arid areas.

Pumping Water to Irrigate... The Wrong Solution for Temperatures

The Get Real Program is focused on having a stable climate system and revitalizing the soil and ecosystems to redistribute ocean water back onto land. The desert reflects solar radiation back to space and has no water vapor to hold heat near the surface at night. This is why pumping water into the desert to irrigate the trees cannot solve for our perceived "global warming" concerns. It is a positive to have additional growth and restore deserts, which may *raise* average temperatures but help *lower* CO2. Many high deserts have temperature extremes, and some desert areas are in northern latitudes with cold average temperatures, which are both more damaging than an average temperature rise overall in deserts. Also, average *air* temperatures are not a big problem for warming the Earth; it is more important to look at heat absorption by dark soot particles on snow and ice that result from coal burning and other sources (more on this later), as well as cloud cover and water vapor scarcity over oceans, which absorb solar radiation. With The Get Real Program, increased water vapor and cloud cover can better balance heat absorption by the white frozen areas, soils, and oceans. This is the key to rebalancing the environment in the future.

More plants mean more transpiration, influencing rainfall and water absorption by the soil. Rainfall falling on vegetation is slowed so that it can soak into the ground. Additionally, roots loosen the soil and create pathways for water to soak in.

There is an incredible myth that the world's average air temperature has something to do with the rate at which glaciers melt when, in fact, it has little to do with it. On the other hand, soot on the snow and ice has an overwhelmingly significant effect on the melt rate compared to the average air temperature rising a degree. The overwhelming heat source is not the air, it is sunlight. Notice the difference between a black car in an open parking lot on a hot summer day and a white car. You can place your hand on the white car easily, but not on the black car! It burns! The temperature of the car is dependent on color and sunlight, not on air temperature. So it is with small particles, including soot on snow. This should demonstrate how important it is where you get your information on problems and solutions for climate change. The Get Real Program is transparent and focused on real impact for our environment for all the right reasons... the replenishment of our environment and food supply in sustainable, doable ways that are not painful to our way of life.

Preserving and Expanding the Amazon

One of the most important efforts of The Program is to provide for the preservation and expansion of the Amazon and other rainforests. Currently, there is no economic value in leaving the rainforest alone, so Get Real advocates for countries to be given real credit for preserving their rainforests, or, even better, even more credit should be given for investing in it by remineralizing it so that rainforests grow better and healthier. Even untouched natural forests can become demineralized by rainfall and leaching; thus, adding non-water-soluble, broad-spectrum minerals can boost growth and sequester more carbon.

Forest fires can't spread if trees are green and thriving. Sadly, at this time of the interglacial period, soil minerals have been leached and depleted in many areas around the world. Areas like the Amazon are not replenished by glaciers, and the rainforest ecosystem often grows on top of a rather depleted soil. When it is cleared for farming, such soil quickly falls off in productivity. The rainforests should not be ignored. There is so much that they can offer to create a healthier planet in ways that cannot be replicated by unnatural solutions. Also, one of the keys is to preserve biodiversity in the rainforest. This should be a high priority within the list of high priorities!

Extraction and Innovation

So many of the forests are so full of deadwood that they emit more carbon than they sequester. Due to the overgrowth and debris of death in the forest, they can be a challenge to manage; however, our forests are a true treasure that can solve for the most challenging of environmental issues of our day, provided we put in the work and investment.

Forests around the globe vary greatly in growth rate and health. Another varying statistic is the number of dead trees in a forest, their state of decay, and whether they are standing or fallen. We have traveled extensively in the U.S. West and have seen from the air and ground how devastated the forests are. It seems that in some cases the death is so widespread and the remaining trees are so at risk, that all threatened trees need to be cut down and trees that are resistant (to insect and other damaging environmental factors) replanted, along with rock dust remineralization for healthy growth. Trees need to be planted according to their adaptation to local factors, including diseases and insects that ravage the region.

When looking at the carbon sequestration opportunity for our trees and large forests overall, a harvested tree that is used for lumber sequesters carbon for a long time while using a tree for biochar sequesters the carbon left after charring for 1,000 years or more. Leaving dead or dying trees to rot or burn releases carbon back into the air quickly. In fact, many forests are so dead that they are net carbon emitters instead of sequestering tons of carbon per acre per year.

Extracting a portion of dead or dying trees from the forest while leaving the healthy trees alone is extremely important. Man's logging practices have been dominated by clear-cutting, the cheapest method of logging. Unfortunately, clear-cutting results in enormous damage to the forest. It often takes decades before optimum growth is achieved in an area that has been clear-cut. In most logging areas, the smaller trees—which are too small to be profitably harvested—are crushed or damaged. Much of the deadwood is not suitable for timber and should be entirely converted to biochar to feed the remineralization efforts. You can read more about this in the Biochar chapter.

Converting deadwood from forests into biochar will drastically improve the water-holding and absorptive ability of soil. Removing wood that is decaying back into CO_2 and sequestering the carbon on a nearly permanent basis will do wonders for the air and soil. Given that one third of the world is forest, we should develop a plan to allocate about $400 billion a year to forest reclamation and replanting worldwide. That will go a long way towards revitalizing the forests and keeping them from being cut down recklessly.

Basalt Rock Dust

Studies, documented in the comprehensive book *Geotherapy,* show sharply higher survival rates and increased growth of trees fertilized with basalt rock dust. There are 750 million acres of forest in America and virtually all of it could benefit from rock dust to boost its health and productivity. Trees need very little in the way of rock minerals to thrive; however, they do require a small amount, and basalt rock dust is proven to dramatically boost growth in depleted soils.

Geotherapy has several case studies demonstrating increases in growth and survival from the use of basalt rock dust. The book is worth reading if you would like more details on rock dust's effectiveness in forests. How to efficiently remineralize large areas of solid timber is a tough question, but like we said, The Program is here for the tough questions.

While basalt rock dust is an incredible solution to many of our environmental challenges, it comes with its own challenges of getting just the right amount to just the right spot at just the right time... not too late, that is. Basalt is one of the most common rocks on Earth, but new mines will have to be opened to increase the supply and shorten transport distances. There is also the question of the means of transportation.

We need to get as much done as possible in the shortest amount of time, while also keeping costs down. Moving a ton or more of rock dust per acre into an intact forest without roads really requires aerial delivery; however, a helicopter is very limited in carrying capacity, and airplanes travel so fast that the rock dust would be difficult to spread evenly. Also, airplane fuel use is high.

A plane flying as slow as 100 mph travels a mile and a half per minute and can be hazardous if it flies low to the ground. An acre of land would be covered in 10 seconds at a 50-foot swath at this velocity. Dispensing rock dust that fast would save time but would not spread it evenly. This is where The Program can step in with innovative solutions... specialized low-speed airships. You can read more on this solution in the next section, Airborne Operation Center.

Most tests with rock dust have involved direct incorporation into the soil or at the planting of seedlings. The finer the rock dust is, the more biologically active it is. Yet, fine powder is liable to drift away in the wind. It is best to dispense the powder close to the ground or at ground level. Specialized low-speed planes, drones, or airships could be built to accomplish the effective dissemination of the rock dust.

The task is large, and transit times would get long given the huge areas and remoteness of forest. Vast tracts of the North American forest that are hilly or mountainous are in bad need of remineralization, so something must be done. We cannot listen to those in lab settings telling us the solutions. We must listen to the farmers and agriculturalists. You can read more on basalt rock dust in the chapter dedicated to it within this book.

Airborne Operation Center

There are hundreds of millions of acres of trackless forest in the United States that contain mostly dead trees. These weak forests are vulnerable to fire as deadwood will easily burn, and there is currently no easy way to remove or use these trees because much of the terrain is very rugged. To preserve the natural landscape and habitats, there are protections against building roads. A novel way is needed to harvest both living and dead trees – one that doesn't involve clear-cutting or major damage to growing trees and that can be managed without disturbing too much of the natural terrain.

One way to remove deadwood and remineralize the forest would be to create airborne operation centers in lighter-than-air crafts that can be anchored over a section of forest. The operation centers would house a few workers, process timber into wood and biochar, and serve as a base for ground operations. There is some helicopter logging of expensive trees, but almost all logging is achieved with logs being loaded onto trucks at a loading point. This type of logging currently requires an access road. There is strong opposition to more roads in wilderness areas, and The Program doesn't advocate for large-scale access road construction in areas without roads.

There are two aircraft types that are lighter than air: hot-air balloons and helium-inflated craft. To turn wood into biochar requires heat. A craft could be built that combines helium or hydrogen inflation with heating from the biochar production to produce much more lift and the ability to support more weight. Imagine a large airship that can lift trees up to it, process the wood for lumber, and convert the rest to biochar to be reincorporated into the forest below. A few workers could either live on it and/or be ferried to it by a drone or helicopter. Please visit www.fullofideas.com for more information on innovative thinking regarding the potential for an airborne operation.

Such a craft would be tethered by cables to the ground and be partially supported by power-generating kites or wind turbines. There is such a density of dead trees that progress would be slow as trees are processed and biochar and rock dust are mixed around the living trees. In fact, in many areas, there is so much deadwood that only a fraction of the biochar would be incorporated into the forest ground; the majority would be transported to desert or agricultural lands that need it. There would be a steady stream of lighter-than-air transport crafts bringing rock dust and taking biochar from the operations center to a distribution base located on the nearest highway or all-weather road. You can learn more about aerial forest management at www.fullofideas.com.

Our modern age has developed high speed aircraft using cheap jet fuel without consideration of the environmental effects. While healthy soil and plants have the ability to sequester far more carbon than man emits, fossil fuel is limited, and we should be focused on long-term sustainability and survival. Remineralizing and revitalizing the Earth is a very large undertaking, and there is not extra fossil fuel capacity to fuel such a project to lower CO_2 to pre-Industrial Age levels, so innovative thinking is key. Lighter-than-air craft inventions are needed!

There really is a need to think much more sustainably about the world's forests so that the next generation has an abundant supply of trees and healthy forests. The focus of the efforts should be on the U.S. national forests since they are being neglected so badly. That's where many of the fires you see on TV spring up. People who have private timberland will manage it with efforts like controlled burns and keep the fuel load down so that their forests don't burn up. They put fire breaks of cleared strips of land in the woods, so there are barriers to a fire spreading. When looking at the U.S. national forests, you have to think differently to address the aversion to introducing roads. This is why it is important to go aerial. States could incorporate aerial concepts for some of the larger state forests.

Call to Action for Our Forest Fires

The Get Real Program knows investment is needed to make the forest a greater carbon sink, as well as investing in wiping out these massive forest fire magnet areas that we see in the news. Year over year, we see the huge fires wiping out thousands of acres of land, not to mention people's homes, lives, and habitats for wildlife. We know what to do and how to do it... we just have to remove the excess fuel from the forest.

There are other staged solutions that can be taken, since removing all of the deadwood would be a massive undertaking over many years. National forests, national parks, and state forests could clear strips of land in the forest to create a grid of safe zones where fires can't spread as easily. This would allow firefighters to put the fire out in a region and for it not to spread to a large area of land. If you did a grid where you had a fire break every mile in forests, that could limit some fires to only 640 acres or less, because once the fire hit the zone that was clear of deadwood, the fire would usually die out. This practice is consistent with forests that are not owned by the government.

THE "WHO PAYS FOR IT?" ... IN CLIMATE CHANGE

People talk about making big changes in society, but they don't often talk about how to pay for it and what the net gain will be. The Get Real Program is specific about the problems, the potential solutions, and how to pay for them. We could have only tax-funded programs, but government is so inefficient and the tax burden on poor people would be crippling, so it is better to have some non-profits do things funded by concerned people and companies. Current proposals such as the Green New Deal ("GND") are dreams for special interests that do nothing to actually lower CO2 levels. Instead, they only slow the rise at an unbelievable cost. The GND proposition offers no way to pay for itself, as if money can just be printed in infinite quantities with no consequences. In the end, there is no "free lunch." The money has to come from somewhere at some point in time, which will end up on our children's backs and their children's backs, and on and on.

It will take lots of money to save the Earth, but the payoff will be huge: a more prosperous and sustainable world that can feed the population well. The technology to remineralize the Earth has already mostly been developed and is available, but it can be improved to make it a more sustainable and energy-efficient process. The goal is a net reduction of CO_2 in the atmosphere without slowing growth. Many scientists are very concerned about the long-term effects of the current CO_2 levels as the climate changes very slowly due to the large thermal inertia of the oceans.

Forests are a real moderating factor in the climate equation, and the die-off of trees hurts the water cycle and the absorption of water by the soil, as well as the carbon cycle and oxygen generation globally. It is grievous to see unhealthy dying or dead forests, and substantial funds are greatly needed to reinvigorate them back to health. It should be clear that dedicated private citizens are accomplishing great things with very little money, while government bureaucracy wastes billions of dollars doing research that goes nowhere.

In the quest to restore the forests, it is vital to support private pioneers fighting the big fight through private global efforts funded by charitable organizations, which will be funded by responsible companies wanting to solve for the real issues of today... instead of promoting only government programs. The Program advocates for letting the more efficient private sector do as much work as possible. Government involvement and funding will most likely be needed on the many acres owned by the government and in certain special heavy-capital areas, but there is so much we can do as private citizens with a cause.

Landowners will be eager to remineralize and reap carbon sequestration dollars by letting the trees grow instead of cutting them quickly. A carbon sequestration plan will need to reflect how long the carbon is sequestered. Technically, many uses of trees really involve long-term sequestration years after the tree is cut.

Newly planted trees don't harvest nearly as much carbon as older trees, but if remineralized at planting, they may actually sequester more than the old demineralized stand did over time. Carbon sequestration payments would be structured to reward long-term sequestration more than short-term that could be reversed with a change in practices. For example, applying basalt rock dust to the land sequesters carbon by chemical action with the rock dust, as well as stimulating soil and plant life to sequester even more carbon. The chemical sequestration is permanent, so it deserves to be paid at a higher price that will cover the cost of spreading the rock dust.

One type of forest that is especially effective in the capture of carbon is the saltwater forest along the coasts. Trees in these forests, such as the mangrove, capture a lot of carbon. Many have been wiped out by dredging and filling up coastal wetlands for construction. Governments or landowners receiving compensation for their carbon sequestering benefits will change the dynamic of these vital areas. In fact, it may be that creating new mangrove forests happen strictly because of the payments.

If the carbon sequestration charitable funding system is global, poor nations with suitable coastlines can reap funds from lowering the world's CO2 level. They could also benefit from improved fisheries and storm protection. Many areas of coastline are rocky and quickly go to deep water with little shallow water in which mangroves thrive. Restoring and creating marine wetlands at the shoreline could be an effective way to earn carbon sequestration dollars while increasing ocean productivity in the area.

One of the easiest ways to increase carbon sequestration and the release of carbon to the air is to pay existing forest owners, whether they are private or government, to maintain and increase their acreage. Currently, there is no economic value, as income, on an acre of rainforest that is undisturbed. Brazil would get substantial revenue from intact rainforest under the Get Real carbon sequestration scheme as the amount of carbon sequestered grows each year.

Some work has been done developing improved varieties of trees and more would be done if trees earned yearly payments for carbon sequestered, with premiums for soil sequestration and long life. The payments for carbon sequestration would be subject to repayment if the carbon is released back into the environment as the goal is long-term lower atmospheric CO_2 levels. Currently, much timber is worth less for sale than getting the annual payments for carbon growth. Timber prices would have to rise to cover the repayment of carbon sequestration payments, which would depend on how the wood was used.

CREATIVE SOLUTIONS FROM LEARNED MINDS

You may wonder what gives us any special insight into trees and forests, as farmers and engineers. For decades, our founder owned property with a commercial timber operation including both personal and contracted logging. He's very aware of the dangers of logging, having had several close calls with potentially fatal accidents. Also, a Get Real engineer has done investment due diligence on forest products companies. We are also well aware of logging's potential to destroy the forest ecosystem, having seen many huge clear-cut areas close-up. We desperately need to change the way we manage and use forests to be sustainable for the future. With remineralization and aerial-selective harvesting, forests will flourish, and the supply of wood will increase to meet the needs of man.

It is so easy to get stuck doing things the same way even though it is fairly obvious that way leads to a desolate future. Those who deny the incredible power of photosynthesis to remove carbon from the air can't see a very happy future. They focus on reducing the output of greenhouse gases while actually doing nothing about it.

The solutions are clear, both the ones in the short term and those requiring innovative efforts for the long term, like airborne solutions that don't currently exist. Converting millions and millions of tons of deadwood into biochar that locks up carbon is a big job, but it's one that has enormous long-term benefits. Planting new seedling trees in rich potting soil of biochar and rock dust, along with some compost or other nutrients, will produce much faster growth and surprisingly higher carbon sequestration. This all requires investment and dedication to get the big jobs done, but first we have to get everyone looking in the right direction… to the trees.

Forests have a big role to play in lowering CO_2 levels and are cost-effective ways to do so. There are more gains from stopping rainforest clearing and burning than other actions, but restoring ocean wetland forests is very close. There are solutions within our reach with the right funding to those charitable organizations already out there doing the hard work. These organizations, like www.getrealaboutclimate.org , just need the funding to create a broader global impact for the benefit of us all. We plan to be part of the solution. Will you?

Man is largely exploiting the woods with little consideration for replacing soil nutrients lost by weathering, erosion, and removal of timber. In many cases, very old growth timber is being cut faster than it is being replanted or the clear-cut area is planted to faster growing, shorter cycle timber. One of the themes of The Get Real Program is thinking of posterity, and we do need to think long-term regarding the forests.

LET'S BE A CHAMPION OF OUR TREES AND FORESTS

The rainforests are the lungs of the world because they release so much oxygen back into the air while sequestering carbon. We must remember that the largest rainforest in the world was once where the Sahara is today. It was lost about 6,000 years ago, perhaps due to the intrusion of early man. If we want to do something about CO2, fixing the forest is one positive way.

With that being the main goal, cleaning up and managing our trees and forests has many other beneficial side effects to our health and way of life. If we move to aerial logging, we can selectively log forests, removing mature trees that are at the end of their life. This will save the younger trees that are coming up from also being destroyed in the process. The innovations discussed here can also expand in scope to address advancements in many other areas of forest life.

Man has a role to be a good steward of the Earth, and that includes being as productive as possible for all of mankind, as well as for wildlife. This includes soil life, which in turn includes our bacteria and all the microorganisms that live in healthy soil that benefit from rock minerals being added. The health of the soil impacts our trees also. We have all been given life on Earth, so we must protect and defend all life on Earth, as we are all part of a large cycle of nature that gives and takes until there is no more.

CHAPTER 8

WONDERS IN OUR WATER: OCEANS, REEFS, WETLANDS, AND MORE

Oceans cover most of the Earth's surface and have a huge effect on weather, climate, and CO2 levels. Unfortunately, the oceans are generally treated with the same lack of respect shown to the soil and trees. Overfishing in many areas and large amounts of destructive pollution are crippling the ocean's ability to sequester carbon. Due to the abuse and overuse, in some areas we are also seeing stocks of fish depleting. This is all very concerning, as the ocean is one of the largest sinks of CO_2 and a major source of our food supply.

Most life in the ocean occurs in relatively shallow coastal waters. These areas are some of the most negatively affected by man's actions. Marine wetland plant life is often being destroyed by habitat destruction (i.e., dredging, filling-in, and poisoning with sewage and chemical runoff). This is concerning, since huge amounts of carbon are used by microorganisms and sea plants in photosynthesis. The phytoplankton that consume CO2 in the near-surface layer of the ocean are impacted in many areas. With a shortage of these specimens to utilize carbon in these areas, CO_2 dissolves in the water, forming carbonic acid, making the ocean more acidic.

Fighting ocean acidity in the regions where reefs grow is crucial. High acidity around ocean reefs is detrimental to coral growth and is a cause of coral mortality. Low pH is caused by high amounts of CO_2 and low amounts of buffering minerals (i.e., calcium, carbonate, magnesium). These elements are the major building blocks of stony coral structures. When these elements are in short supply, corals suffer slow growth rates or death.

A dead reef results in total loss of all ocean life in the area; fish and all other inhabitants (i.e., crabs, snails, starfish, and many other bottom-dwellers) die off, or if lucky, swim off to a living reef. When the die-off starts on a reef, stored carbon, nitrogen, phosphate, and other elements in the coral and its dependents are released to the water. This immediate large dose of elements compounds the problem by over-feeding competing algae that can end up covering reef rocks and coral, smothering what life remains. If there is a significant die-off, an underwater forest of lifeless calcium and carbonate skeletons is all that's left.

Alternatively, healthy coastal and tidal areas have an amazing ability to sequester carbon, and shallow systems provide many more benefits than just carbon sequestration, one of which is serving as nurseries for the fish and sea life that often make up the basis of ocean life's food chain. We all will suffer if the food chain suffers in the ocean.

Because the ocean covers most of the Earth and currently holds the vast majority of CO_2, it is vital that it be part of the solution to lowering CO_2 levels in the atmosphere. Unfortunately, with the oceans, reefs, and wetlands suffering, investment is needed to make the most of this natural carbon sequestration. As with the soil and the trees, farmers know how critical remineralization is to sustaining life and the health of our environment in order to get the most out of our natural resources. This holds true for our oceans and wetlands too.

The following is a selection from the abstract of a paper on carbon sequestration by Thomas Goreau from the Soil Carbon Alliance website at www.soilcarbonalliance.org. It should be noted that Dr. Goreau doesn't endorse some of the ideas proposed in this book, such as floating islands, but he is an authority on near-coastal marine life and has proven ways to improve ocean life affordably.

> "Marine wetland soils (salt marsh, sea grass, mangroves) occupy less than 1% of the Earth surface, but hold about a half of wetland carbon and a quarter of global soil carbon (more than the atmosphere or biosphere), and account for about half the carbon deposition in the ocean. Newly developed methods now allow rapid regeneration of marine wetland soils and carbon storage, which will be one of the most effective soil carbon sinks for the cost and area required, while providing valuable benefits for shore protection and fisheries habitat."[17]

Based on Dr. Goreau's work it can be seen that marine wetlands, mangroves, salt marshes, and sea grasses, covering around 1% of Earth's surface hold more carbon than the atmosphere, burying around half the carbon in the ocean. Also, some marine wetlands are rapidly vanishing ecosystems and their restoration could provide large carbon sinks in the small areas for reasonable costs, restoring critical fish nurseries and protecting coastlines from erosion. Some efforts to restore marine peat soils fail because new plants wash away before their roots can grow. These problems may be overcome with biorock electric stimulation methods, allowing sea grass, salt marsh, and mangroves to be grown where all other methods fail. This would expanding carbon-rich ecosystems seawards where in many cases they are rapidly eroding away.

Improving the health and increasing the area of near-shore wetlands, both salt water and fresh water, is vital to sequestering carbon and

[17] https://www.soilcarbonalliance.org/2017/02/01/regenerative-development-to-reverse-climate-change-quantity-and-quality-of-soil-carbon-sequestration-control-rates-of-co2-and-climate-stabilization-at-safe-levels/

has the added benefit of a potential harvest of biomass to digest for biogas. The Program shares more on biogas in a later chapter.

Bold thinking and new methods are needed to restore the oceans to a better condition. The following is a summary of the scale of ocean life sequestering carbon and some of the key nutrients that limit its activity. It should be noted that it is very difficult to attribute the need for very low amounts of trace elements in either soil or seawater. Quoted from an article titled "Nutrients that Limit Growth in the Ocean" by Bristow, et.al. in *Current Biology.*

Phytoplankton form the basis of the marine food web and are responsible for approximately half of global carbon dioxide (CO_2) fixation (⊠ 50 Pg of carbon per year). Thus, these microscopic, photosynthetic organisms are vital in controlling the atmospheric CO_2 concentration and Earth's climate. Phytoplankton are dependent on sunlight and their CO_2-fixation activity is therefore restricted to the upper, sunlit surface ocean (that is, the euphotic zone). CO_2 usually does not limit phytoplankton growth due to its high concentration in seawater. However, the vast majority of oceanic surface waters are depleted in inorganic nitrogen, phosphorus, iron and/or silica; nutrients that limit primary production in the ocean.[18]

REMINERALIZATION IS REQUIRED

Scientists are vague regarding the role of trace elements in the ocean, but they speak fairly consistently about iron deficiency. Here is a conclusion from a paper on trace elements in the ocean called "Diagnosing Oceanic Nutrient Deficiency" by C. Mark Moore:

> Analysis of the data contained within the GEOTRACES IDP2014 confirms that the open ocean, or at least the

[18] Bristow, L. A., Mohr, W., Ahmerkamp, S. and Kuypers, M. M. M. (2017). ***Nutrients that limit growth in the ocean, CURRENT BIOLOGY***, 27(11), R474-R478, doi:10.1016/j.cub.2017.03.030.

> Atlantic, is stoichiometrically deficient in Fe [iron] everywhere outside of the macronutrient (N,P) [nitrogen, phosphorus] depleted surface (sub-) tropical waters. Such a conclusion appears to be reasonably robust to caveats associated with the plasticity of phytoplankton Fe:N ratios.[19]

Researchers and scientists are somewhat divided on the desirability of adding nutrients to the ocean. The Get Real Program sees the potential and is working hard to prove multiple concepts to support great ideas for our oceans. There is so much time and investment being put towards the wrong solutions or limited impact for our climate and the overall health of our world. This is why we were compelled to write this book to bring about awareness of the wrong messages and get us reset towards real impact that we can see in our lifetimes.

Shallow areas, which are the most productive, are often badly demineralized. Every pound of fish that is removed from the ocean also removes from the water a small amount of elements stored in the fish (i.e., carbon, calcium, phosphate, and trace elements). Studies need to be done on different ways to remineralize the ocean directly. Remineralize.org would be the perfect body to do more studies with its historic database and depth of expertise. Oh, there is more, too…

Let's Eat!

Remineralizing the land and sea is key. Fish are different from warm-blooded animals in that they can subsist on very little food as they produce no heat to stay warm. Yet without ample nutrition, fish don't grow. In many places in the ocean there is a shortage of food, keeping fish from growing as much as they could. A remineralized

[19] Moore, Mark C, ***Diagnosing Oceanic Nutrient Deficiency***, The Royal Society 374.2081, 28 November 2016. https://royalsocietypublishing.org/doi/full/10.1098/rsta.2015.0290

stretch of ocean can produce a lot more phytoplankton to feed the next step in the ocean food chain. A lot more fish are needed to feed the world sustainably.

The ocean is so large that our efforts to remineralize it will cover only a tiny fraction of it, even after years of work. Nonetheless, it will make a difference. All we hear is talk, talk, talk. All we see is money floating out to areas where you never hear of any real impact or change. Something needs to happen now, in order for us to save the world before we experience the impact of our depleting oceans, reefs, and more. Let's make real change happen together. Let's focus on the areas of real need and impact, which is increased productivity in the ocean to feed people and sequester carbon.

CRITICAL AND INTERESTING SOLUTIONS FOR OUR ENVIRONMENT AND OUR WATER

Rock Dust to the Rescue AGAIN

Growing acidity and a lack of rock nutrients in the ocean harm all types of ocean life, from coral reefs to large sea life. There is a need to remineralize the ocean waters with various rock dusts that are optimal for improved conditions in the oceans. Massive applications of basalt rock dust to the land will also result in some of the elements leaching into runoff and traveling to the ocean.

Think of the millions of tons of microscopic rock dust that washed into the oceans during the long glacial period. It is important that the nutrients in the rock dust be quickly available to ocean life, without waiting for another natural phenomenon. Microorganisms that can break down minerals bound in rock dust are specialized and may not exist in many shallow water areas where oceanic nutrients are in short supply. This is where we can make an impact.

Sunlight is critical for photosynthesis in the sea, so it would be best not to have the rock dust cloud the water, making it harder for sunlight to penetrate. It is possible that the rock dust will be pneumatically sized with only the smallest rock dust mixed into the water as it would be very difficult to get it to settle on the ground and not blow away many miles in the wind with land application. Since ocean water is not as densely populated with organisms as healthy soil is, it is important for the rock dust to be extremely fine. This makes it readily available and keeps it in suspension in the top layer of water where it can feed phytoplankton and other organisms that are the base of the food chain.

Pulverizing basalt rock into dust at sea offers the opportunity to segregate the produced powder by size. It could then be distributed with the very smallest going into the sea, where it would remain suspended for a long time to feed microorganisms like phytoplankton. When crushing rock particles, the rock fines that are smaller than the desired size are normally considered waste. Very small particles that won't easily separate out of the air except by rain are difficult to use on land.

A vertical shaft impact rock crusher can make very fine rock particles. The crusher's output could be run through a pneumatic cyclone where larger particles would be captured and only the very smallest remain in the airstream. These could then be blown into the water at its surface. These microscopic rock particles would float away in the air for long distances, like dust from the Sahara Desert that travels all the way across the Atlantic. Blowing the particles into the water would allow them to be suspended near the surface for a long time and serve as a source of nutrients for the web of life in the ocean.

As far as putting out rock dust, it should be clear that we are not talking about a thick cloud of rock dust in the water. The distributed amount would be very small. A vessel traveling 10 miles an hour and grinding 100 tons of rock an hour will grind a ship load of 100,000 tons in six weeks at sea. It may very well be that the ship offloads most of its rock dust for land application, but for this example, let's say that five percent is mixed into the top layer of seawater. Assuming the rock dust is spread over a 500-foot-wide swath, the result is less than 20 pounds of rock dust per acre.

Unlike the land, where there have been many studies on the effects of basalt rock dust, there is less research on the ocean – perhaps because it is held in common and is so vast. The ocean will play a big role in transporting the extremely large volumes of rock that are needed to remineralize large amounts of distant land masses capable of plant growth. Ocean transportation, combined with capturing wind, solar, and wave energy to produce rock dust, can be a very efficient way to remineralize the ocean and deliver rock dust to faraway lands. However, more research is needed on the effect of basalt rock dust microparticles on phytoplankton growth before it is done at sea. Perhaps there is a better mix of rocks or minerals. The good thing about basalt is that there is a lot of it spread around the world. This is critical because the need for minerals is so large on land and sea.

Biochar Has More Than One Use

Chemical runoff from conventional agriculture and sewage plants causes nutrient overload and toxic algae blooms that kill sea life. By using biochar on the land to sequester nutrients, the large-scale wasting of geologic plant nutrients and nitrogen will be sharply reduced.

Conventional Practices Poisoning Our Waterways

The discharge of sewage effluent into waterways is another area that cannot be ignored as a challenge in keeping our oceans healthy. The enormous size of the oceans makes people think it is okay to dump waste into them, but lasting harm is done to coastal areas as much of the effluent stays close to shore in vital wetland growing areas. This is an area where the government should take increased action to protect and preserve our waterways.

Floating Lagoon Islands

The majority of the ocean is very low in life, especially the deep ocean far from land. Creating artificial reefs on floating lagoon islands would boost sea life in areas that are currently nearly devoid of life, without affecting existing coastal areas. Nutrient-rich deep ocean water could be pumped up to the surface by thermal siphons to provide for the sea life in the lagoons. We have all likely heard of the amazing fishing around offshore drilling platforms. This is because the platforms are habitats for sea life, as much sea life depends on having a structure to cling to.

Deep-sea fish farms are advantageous because they don't involve destroying coastal habitat. Small, floating island communities could produce food and fuel as well as sequester a lot of carbon. An anchored vessel could manage the fish farms and process the fish for shipment. Lots of carbon would be sequestered by the sea grass and other saltwater plants such as seaweed/kelp. The whole enterprise could be powered by the sun, wind, and waves with a backup power source such as natural gas-fired turbines.

Another thing that offshore floating lagoons can accomplish is growing large quantities of aquatic plant life for biogas production at the co-located biogas digester. You can read more about this and other fascinating inventions in the smart energy options chapter. The idea for floating islands is certainly a bold one, so we look forward to producing more research in this area as part of The Get Real Alliance's efforts for a more prosperous world. It may not be feasible today, but if accomplished with advanced technology, it could drastically increase sea life and carbon sequestration.

Using Wave Energy for Power

There is great energy in ocean waves. Wave energy can provide lots of power and has incredible potential for more advances. Unlike solar, wave energy can provide 24/7 power. There is a need for extensive research on improved methods of capturing wave energy. Massive floating islands may be able to serve as wave energy capture structures, transmitting power back to the shore through underwater cable or used on site in the case of structures such as lagoons far from land. To help feed humanity sustainably, the ocean needs to be much more productive.

If humanity is to power itself from energy derived from the sun, a new way of doing things is necessary. While they work, we don't need to go back to traditional wind-powered sailing craft designs. Better ways just may be possible. A large craft, powered by sails and wind turbines combined with wave energy capture and solar power, should be able to achieve higher speed and smoother travel. In traditional hull designs, the craft of any size rocks in the waves, oftentimes heavily. A wave energy-capturing craft should be able to knock down the waves as they approach the craft. See the founder's invention company www.fullofideas.com for updates, new ideas, and additional information.

The Get Real Program is passionate about new, innovative ways of transportation and sustaining our energy needs while solving for environmental concerns. More information is shared in the chapter on smart energy options.

Wind-Powered Ships

There is a real need to fuel ships with natural gas to lower emissions, but an even better solution would be a return to wind-powered ships in a "hybrid" form. One thing that hasn't been explored is a ship powered by both wind and waves. There are some cargoes that are time-dependent and need rapid transit, but many cargoes can take extra time to get to their destination. This would include the large amounts of basalt rock dust needed to remineralize the Earth and to cause chemical sequestration as the rock breaks down.

As mentioned previously, basalt is a very common rock, but there are areas where there is not a ready supply. There is not enough extra oil and gas production to fuel a major remineralization effort, and it is counterproductive as far as reducing atmospheric CO_2 levels to use lots of extra oil and gas to mine, transport, and spread rock dust when renewable power sources could be used. Though it may seem far-fetched, a wind-wave-powered ship could actually grind rocks into dust as it travels across the ocean. As mentioned, there is more on this topic in the smart energy options chapter.

Re-imagining our Harbors

Another traditional way of doing things that will need to be substantially restricted is the stationary land harbor. The traditional, narrow, long, and relatively shallow draft boat can go up a river or inlet that is the basis for many harbors. Traditional harbors shelter the watercraft from waves with rigid barriers that the waves bounce off, causing them to dissipate. Lots of wave energy is wasted bouncing against the shore or harbor breakwaters.

An example of a more innovative idea for today is an offshore floating harbor for the many areas that don't have a natural harbor. This would allow transport ships to dock in a calm harbor while capturing wind and wave energy for the harbor and docked ships. These harbors would increase the use of ships for transport, which is energy-efficient even if the ships run on fossil fuel and could be carbon-neutral if wind and waves are used.

Most areas of the world can't use ocean shipping because they don't have a harbor. Many countries have shorelines but no major harbor. Transport of large quantities of rock dust around the world will require new watercraft and a new way of unloading the contents near shore.

In addition, large amounts of power can be gathered from the offshore waves by an offshore docking port as seen in U.S. Patent #8,858,149 on the Full of Ideas website. Large offshore docking ports can be positioned off any coast to provide a place for vessels to unload for transport to shore by smaller vessels or barges that can access limited ports. In Africa and Latin America, where roads are limited and ports sparse, it may be that lighter-than-air transport craft are needed to move cargo and rock dust inland effectively.

Ocean Plants Can Help

Producing biogas from ocean plants is also another feasible option. This can be done near the shore if "seaweed" was promoted through remineralization and the creation of environments in which it can thrive. Underwater structures where plants can attach in areas of shoreline that have deep water just offshore can be easily constructed. Also, today, lots of sea plants just wash up on beaches to rot or go to the landfill.

Researchers do work in the areas that have funds available, no matter how important or vital to humanity. Plant breeding focuses on areas where people will pay for improved plants. Developing varieties of sea plants that offer superior carbon sequestration and/or food or fuel production can help make the world a better place; thus, research dollars should be dedicated to this area.

Increasing the number of sea plants will help lower ocean CO_2 levels, which cause acidity. A lower pH in the ocean has negative effects on coral, along with higher temperatures. A small amount of any carbon tax and charitable contributions is needed to fund research and ocean plant development to increase carbon capture and sequestration.

The Big Cleanup

Part of the solution is avoiding things like throwing plastic into the waterways, but positive action is also needed to rebuild ocean environments. Presently, the number of water bottles created in a year is unbelievable. These bottles have to go somewhere as they are used, but trashing the ocean is not the answer. At the same time, we need to restore and increase nature's filtration systems of shellfish, marine wetlands, and forests such as mangroves.

IMAGINE IT... CARBON SEQUESTRATION ADVANCES WITH BIOROCK

There are real carbon sequestration solutions for the oceans, reefs, and marine wetlands. The best solution is large-scale rejuvenation of reefs with biorock, but there are so many more advances for biorock. Biorock is a name for electrically conductive rock deposited in shallow seawater. Biorock involves the use of conductive steel lattices to create very low-power electric fields generated between an anode and a cathode to greatly stimulate marine growth of all kinds. Biorock has been around since 1979 and has been used around the world on a small scale with great success. The ocean needs a massive deployment of biorock to restore reefs and marine wetlands.

The use of existing technology, such as biorock, needs to become widespread if coral reefs and shoreline wetland habitats are to be rebuilt. Phytoplankton need minerals to thrive, and they can sequester a lot more carbon if fed well. Unfortunately, most climate activists focus on individual localized problems and don't give larger biologic processes much notice. They are missing out on positive solutions to the issues. Historically, we have essentially been mining the sea and now it is time to invest more in rebuilding sea life.

Using biorock to rebuild oyster beds and clean the ocean water could help reverse the recent historic trend. With positive solutions, such as the use of biorock and rock dust, we can stop mining species out of existence and see the ocean produce sustainably with rising productivity. The very thorough book *Innovative Methods of Marine Ecosystem Restoration*, by Thomas J. Goreau, includes a number of papers that validate successful application of biorock with a variety of species. The paper entitled "Reef Restoration Using Seawater Electrolysis in Jamaica" says the following about the fundamentals of the process:

> "A novel technology developed by architect Wolf Hilbertz in the 1970s uses electrolysis of seawater to precipitate calcium and magnesium minerals to 'grow' a crystalline coating over artificial structures to make construction materials... Electrolysis of seawater results in mineral deposition at the cathode. Typically, the cathode is built out of expanded steel mesh or other steel structure. The applied voltage is harmless but changes sea life attached and near the cathode of the system. The electrochemical reaction prevents corrosion of the metallic cathode as it is plated with sea minerals that can have a strength exceeding concrete."[20]

The following is the abstract from a presentation given at the 2018 International Summit on Fisheries & Aquaculture.

> "Biorock mariculture technology is a novel application of marine electrolysis, which grows solid limestone reefs of any size or shape in seawater, that get stronger with age and are self-repairing. Biorock reefs can be designed to provide habitat specific to needs of hard and soft corals, sponges, seagrass, fishes, lobsters, oysters, giant clams, sea cucumbers, mussels, and other marine organisms of economic value, or grow back severely eroded beaches at record rates. Biorock reefs, and surrounding areas, have greatly increased settlement, growth rate, survival, and resistance to severe environmental stress from high temperatures, sedimentation, and pollution for all marine organisms observed. This allows marine ecosystems to survive otherwise lethal conditions

[20] Edited by Goreau, Thomas J., Trench, Robert Kent, ***Innovative Methods of Marine Ecosystem Restoration***, CRC Press, Taylor & Francis Group, Boca Raton, FL, 2013.

and be regenerated at record rates even in places with no natural recovery. These remarkable findings seem to result from weak electrical fields poising the membrane voltage gradients all forms of life use to generate biochemical energy (ATP and NADP), causing enhanced growth of all species. Biorock technology provides a new paradigm for whole-ecosystem sustainable mariculture that generates its own food supplies, the antithesis of conventional mono-species mariculture dependent on external food inputs, whose wastes cause eutrophication that kills off surrounding subsistence fisheries. Potential applications include fish, crustacean, and bivalve mariculture, algae mariculture, pharmaceutical producing species, and floating reefs for pelagic fishes. The power requirements are small and can be provided by solar, wind, ocean current, and wave energy. The techniques are ideally suited for community-managed mariculture, if investment funding were available to subsistence fishing communities."[21]

The full presentation can be seen at the Global Reef Alliance website along with lots of other quality material. Please see www.globalcoral.org , if you are unconvinced by this brief chapter.[22]

The world's reefs and shallow water systems that are in decline can be rescued with large-scale use of biorock and rock dust. Biorock and rock dust can make them healthy and sequester carbon. The Get Real Program is a supporter of the Global Reef

[21] Goreau, Thomas J.F., ***Biorock Technology: A Novel Tool for Large-Scale Whole-Ecosystems Sustainable Mariculture Using Direct Biophysical Stimulation of Marine Organism's Biochemical Energy Metabolism***, http://www.globalcoral.org/2018-international-summit-on-fisheries-aquaculture/

[22] https://globalcoral.org/_oldgcra/reef_restoration_using_seawater.htm

Alliance and their use of biorock around the world. It must be remembered that living things all sequester carbon. This focus on reefs and shallow coastal areas, in conjunction with saltwater marsh and mangrove forests, can sequester a lot more carbon. These areas are the hosts for so much carbon-based life. Research is needed on the optimum way to remineralize the ocean and shoreline.

The biorock process is amazing for boosting coral growth and other marine life. Utilizing it for our innovative advancement of our environment needs to be done on a massive scale over hundreds or thousands of square miles! See www.globalcoral.org or www.biorock.org for more information on the great process of creating man-made reefs and other structures underwater. It should be clear by now that there are numerous positive solutions to the many environmental issues facing society.

OPPOSING VIEWS

There will be opposition to remineralizing the ocean from several fronts. There are some who oppose any efforts to modify the ocean and are opposed to man's interaction with nature, whether good or bad. Then there are those who don't believe that natural biological processes can really lower CO_2 levels dramatically.

There is a gloom and doom mindset that advocates for man to lead an austere life. Following this path wouldn't lower atmospheric carbon levels; it would only change modern life for the worse. There are some who don't want to see a positive solution enacted that doesn't entail great sacrifice.

There will also be a debate on how to spend carbon sequestration dollars most effectively. But first, the possibility that biologic processes can really do the job has to be endorsed by people who currently largely dismiss it.

THE "*HOW*" BEHIND THE WHAT WITHIN OCEAN SOLUTIONS

Because coastal waters are under the sole control of local governments, programs to enhance carbon sequestration and remineralize our waters will include red tape and much supervision in one way or another. The Get Real Program prefers non-governmental solutions, but property rights stop at the waterline. We must create an effective public-private partnership to create the greatest impact.

Even with the governmental control, you may wonder how we could possibly pay for such large ventures. The answer is that it would be easy if society truly decides to remove more carbon from the air than is emitted and actually lower the CO_2 level to an optimal one. There will be a more thorough discussion of how to fund a prosperous sustainable world in later chapters, but for now, know there are solutions.

A theme throughout this book is that excess CO_2 in the atmosphere is a symptom of a sick world. It will no longer be a problem if the world's soil and its oceans are restored to health. Many of you who read this book may not be convinced that rising CO_2 levels are a problem. Unlike many activists, The Get Real Program is not saying whether the CO_2 rise is a problem or not; it is just a stated fact that the levels are rising, whether you consider that good or bad.

The things that are making levels rise, such as soil destruction and demineralized land and sea, are actually huge problems. These challenges to the Earth's sustainability are the focus of The Program. By fixing these ecosystems, the CO_2 issue can be resolved and all while producing larger quantities of healthier food, creating better habitats for wildlife, and making a more beautiful world!

Either way, there is also a need for creativity and invention to move new technologies into use to advance society. There needs to be a workable economic incentive to move away from the use of geologic oil and gas when renewable sources are available. The

Program advocates for shifts over time, rather than an abrupt halt to current energy sourcing that we sometimes see advocated by our political leaders. With The Get Real Program, change is incentivized over a timetable that is workable and more palatable for everyone. You can visit www.fullofideas.com for more inventions that can make a difference.

Even with substantial use of renewable energy to mine, transport, and distribute rock dust, fossil fuels will be needed to do some of the work. The Program always offers the solutions with the problems, unlike many activists. Not only are there solutions, but it is also important to see more investment in research for advances to what we know, what we can do, and how it is done. We implore you to get involved and invest your time and be part of the "what and how" for action to happen now, before overall natural resources are so depleted and degraded that change becomes more of an uphill battle than we already face.

A NEW NAUTICAL TOMORROW

A good bit of time has been spent in this chapter addressing possible new ways to travel, capture energy, transport products, capture carbon, and revitalize the ocean and shoreline with the goal of improving the climate, the ocean, and life in the ocean. If we are going to have a prosperous and sustainable future, we need to do things differently. The inventions and methods described in this book may or may not be immediate practical solutions, but they are examples of new thinking and a catalyst to start the conversation.

One thing is clear: going fast uses lots of fossil fuel whether on land, sea, or air. The fossil fuel era, and the oil era in particular, marked a huge increase in the speed and volume of travel. While GRA proposes that CO_2 levels can be lowered without giving up fossil fuels, we realize that we need to conserve them and learn to live without using them so much. Whether we have 50, 60, or 100 years' worth of geologic fossil fuels left at the current world consumption rate, we need to start thinking of how to live without them because it's not a function of *if* but *when* they will run out.

So many of Earth's resources are running out on a similar timeline. Mankind has gotten away from thinking about posterity, living so that our great-grandchildren can have a good life, and instead is behaving in ways that put future generations in peril. For instance, putting eight million tons of plastic into the ocean each year and overfishing are behaviors that don't portend a promising future. We need to focus on making things better over time.

The following is another passage from Goreau's website, globalcoral.org. Goreau is a leading proponent of using the biosphere to lower CO_2 levels instead of the futile focus on just slightly reducing carbon emissions by man.

> Two strategies maximize soil carbon sequestration cost-effectiveness: biochar and marine wetlands: biochar carbon is stable for thousands to millions of years, whereas marine wetland peats are the richest soils in carbon because lack of oxygen prevents decomposition. Regeneration of mangroves, sea grass, and salt marsh peats could sequester the needed carbon in around 1% of the Earth's surface.

You have already helped fund *Remineralize the Earth* through the purchase of this book, but if you like the idea of actually remineralizing the Earth, doing so requires more support to fight the negative mindset of the mainstream climate community. The Get Real Program must invest dollars where they will remove the most carbon from the air with a higher value placed on permanent removal.

One of the many things a dollar donated to The Get Real Alliance does is lobby for improved treatment of marine wetlands along with organizations such as www.globalcoral.org. The Get Real Program sees restoration of the marine wetlands using biorock as the only proven way to sequester CO_2 on a large scale in the oceans. The Get Real Program is sounding the blow horn for rapid research and development on the topic of wetlands rescue. Your support is needed at any and all levels. Giving to www.getrealalliance.org is broad-based support but if you itemize your donations you need to know that as an advocacy group, donations to it are not deductible. Direct donations to the mentioned charities are deductible.

The efforts will result in an increase in ocean carbon sequestration by increasing the activity of phytoplankton and marine wetlands as well as restoring and creating reefs that harbor so much life in the ocean. Giving money to www.globalcoral.org promotes the further application of biorock and other proven ways to bring life to the oceans.

Our goals for land, ocean, and life on Earth are not modest, but with everything that's at stake, bold action is called for... and now! Many local ecosystems are on the verge of collapse. Some have already collapsed, such as the areas of ocean that are nearly devoid of fish due to overfishing and reef destruction. Many reefs are dying. There is a desperate need for drastic action to save the oceans, before we also need saving from the subsequent impacts.

CHAPTER 9

LET'S GET REAL WITH SMART ENERGY OPTIONS

Renewable energy is the shiny new thing that everyone is chasing. Unfortunately, some solutions under the renewable energy umbrella are not worth chasing and others are downright fraudulent. Dollars are being funneled into project after project under the guise of being a miracle problem-solver for climate change, whether there is a proven potential for impact or not. The Get Real Program is attempting to break it all down here to help you understand what investments are worth it and which ones are just chasing dreams that will never be realized.

The Get Real Program believes in progress and understands that we don't need to take rash action on energy. We need sustainability while also phasing out the majority of coal in favor of natural gas generation. The achievable programs outlined as part of The Get Real Program would lower CO_2 levels while still using oil and natural gas for a long time and ongoing as primary backup. We should start our energy discussions from a realistic place and not expect miraculous results out of our energy sources.

Change is needed, and the time is NOW! Fossil fuel reserves are not infinite, so we need to work towards smart energy systems that return many times the energy needed to create them and that last for a long time. There is a smart, realistic way to do this, and The Program can guide direction and investment from insight based upon experience. This experience will help get to the core of the issues with minimal waste, unlike so much of what we hear about that is being funded today. We say, follow the dollar and you will learn much and see a scary reality!

A PLACE IN TIME

Just because something has worked in the past does not mean it was the best way to do things or that it can continue indefinitely into the future, and the same is true with energy. Earth's geologic fuel resources are a little like a limited supply of food on a long voyage (like a mission to Mars!). Planning and developing astutely and correctly are key.

Nature's Energy Connection

At the center of our solar system resides a massive ball of hot plasma—a natural fusion reactor—whose light is behind all life on Earth. Photosynthesis evolved relatively early in life on this planet, converting the sun's abundant luminous energy into chemical energy. For most of human history, people have relied on the derivatives of photosynthesis. We consume plants, animals that consume plants, and animals that consume animals that consume plants.

Over many, many hundreds of millions of years, plant life—first single-celled organisms and later multicellular ones—thrived on the warmth and energy of our yellow star. These organisms converted to biomatter stored below the Earth's surface, which, with enough time and pressure, would eventually transform into fossil fuels.

We have also used wood as a fuel source since the discovery of fire. We continued to evolve our energy resources and have rapidly industrialized in more recent centuries with the use of fossil fuels. From coal-fired steam engines to internal combustion-powered automobiles to jet-fueled aircraft, modernity is tied to those photosynthetic processes that took place over countless eons.

Progress Can Come at an Unwelcome Cost

Damage has been done in our quest for energy resources and other modern practices. Large forests have been cut down too fast, in a non-renewable manner. Many scientists agree that a rainforest once stood in the location now dominated by the huge Sahara Desert. It is notable, however, that man's use of fossil fuel has resulted in reforestation in many areas too. This is because fossil fuel now provides energy that was once provided by wood; however, our reliance on cheap fossil fuels has steadily depleted reserves. This is only one area where we need change for sustainability and so much more.

Let's Introduce Fossil Fuels – Our Primary Source of Energy

Fossil fuel, or hydrocarbon, is made up of both carbon and hydrogen. Fossil fuels range from methane, which has one carbon atom and four hydrogen atoms, to coal that is nearly all carbon. When we look at the tons of CO_2 emitted, it is obvious that it would be better to burn something that gives off a lot of energy from hydrogen, such as methane, rather than burning coal which is so heavy in carbon. However, and as important as they may be, fossil fuels are also ultimately a finite resource.

If we rely entirely on a finite resource and deplete it, then it will eventually run out. Our reliance on fossil fuels has steadily depleted reserves, and we have moved from sources that are easy to access to sources that are more difficult to access. We must be long-term focused, thinking not in terms of having sufficient geologic resources for decades, but for centuries.

Technological innovation in recent years (i.e. horizontal drilling and hydraulic fracturing) have rendered false the predictions of an impending peak in oil and gas production, but there is still a limited amount. Today, many feel that the environmental threat caused by burning fossil fuel and releasing CO_2 is serious, justifying radical action to slow the rise of atmospheric CO_2 at any cost, no matter how disruptive. The Get Real Program sees a less painful pathway to energy production becoming sustainable and maintaining national security, and all while being less damaging to the environment.

It Comes Down to Cost

It does not make sense to waste a vital, but limited, resource just because we can acquire it at a low cost in the present. It is crazy to base the price of a commodity on what it costs to produce, assigning it no intrinsic value. We put no value on what it would take to create oil or natural gas, just what it costs to extract it. Nature must be preserved, but we must also use what we have been given naturally for our own existence and prosperity.

The extractor is happy getting more for a product than the cost of extraction. However, we now put little value on the extreme time and pressure it takes for biomatter created from the sun's energy to become petrochemical resources over many, many millions of years. This process cannot be easily or cheaply replicated by man. The extractor earns a margin on this product – a product that he or she could never replicate or create independently. Super cheap fossil fuel prices hamper long-term energy thinking. Fortunately, there are other options.

OPTIONS FOR CHANGE

Nuclear

If we are going to get real about renewable and low carbon emission energy, we have to look realistically at the full range of options. If our goal is to produce energy with low or zero carbon emissions, nuclear power needs consideration from a realistic perspective. We certainly shouldn't abandon existing powerplants that have decades of low-cost life in them out of misguided concerns, or worse, risking a less stable power system.

Now, nuclear energy is not renewable, but it is carbon-free, and some advocate for more nuclear. Unlike many, we believe there is a role for nuclear, and we should not prematurely close nuclear power plants. Michael Shellenburger, in his very good book *Apocalypse Never* makes a long, strong case that there are ulterior motives behind some anti-nuclear activists, including a desire by some to see man go back to a much humbler way of life of scarcity. Some of the opposition has been funded by fossil fuel companies that didn't want to lose the market for electricity. It is important to follow the dollars![23]

Uranium is not an infinite resource, and current reactor designs are wasteful of nuclear fuel. The thermal efficiency of a nuclear plant is low because the water used to extract heat from the reactor must be kept below its critical point, which is too cool for the top efficiency of a steam turbine. So, most of the heat given off by the uranium is waste heat that ends up heating a body of water or air to no gain. Then, the uranium deteriorates and doesn't attain a sufficiently high temperature to make the steam. So, it is pulled out of the reactor and continues to decay in a holding pond that has to be cooled, because the fuel is still giving off a lot of heat.

The only way to effectively use uranium would be in small safer reactors located near cities where all the waste heat could be used for other purposes, but this is currently unlikely because of excessive concerns about radiation leaks. The Program only sees nuclear energy as a viable long-term energy source if there are substantial improvements such as some new modular designs that appear promising – these new modular plants are much less susceptible to meltdown and appear much more efficient. Nuclear power is currently a low percentage of the world's electric generation, and there is estimated to be a 130-year supply at the current rate of use and pricing; there is much more at higher pricing. If we build a lot more plants, the reserves would be used up far faster. At higher

[23] Shellenburger, Michael, ***Apocalypse Never: Why Environmental Alarmism Hurts Us All***, Harper, illustrated ed., June 30, 2020.

prices, it is possible to extract uranium from seawater and have a 30,000-year supply.

Much smaller reactors are safer because there is less material to melt together, and it is also easier to provide enough cooling to prevent a meltdown. The problem has been the utility industry designed around huge remote power plants and long-distance transmission wire wasting lots of the energy as waste heat and powerline losses. There are rumors of one early nuclear researcher having a tiny amount of radioactive material in a converted wood stove in his office to provide clean, safe heat. See www.environmentalprogress.org for more information on nuclear energy. Go to www.fullofideas.com for more updates on novel and better nuclear designs.

We need to keep existing nuclear plants in operation even if they need some subsidy. The only really safe nuclear heating is using the heat of the Earth, which is from nuclear decay. Geothermal heat energy is many times greater than the heat from the small amount of uranium found in the Earth near the surface. And, there's more on energy…

Wind

Wind energy also gets a lot of attention with little recognition of its variable nature and occasional outages due to either low wind or too-high wind speeds. There are many problems with wind energy. No one likes to talk about how wind turbines need to have wind at a given speed range to work—not too fast and not too slow—and about the difficult disposal issue for used components and the huge blades.

Turbines are subject to icing in cold weather, and data is needed on how turbines hold up in hurricanes or big windstorms. Durability, longevity, and end of life recycling costs are often overlooked. A 20-year life sounds like a long time, but time flies by making total costs important. *Bloomberg Green* ran an article about the difficulties of disposing of wind turbine blades. Some blades are being torn down after only 10 years of operation due to being uneconomic compared to new models and the higher cost of operation due to maintenance. The following is a selection from the article:

> "Tens of thousands of aging blades are coming down from steel towers around the world, and most have nowhere to go but landfills. In the U.S. alone, about 8,000 will be removed in each of the next four years. Europe, which has been dealing with the problem longer, has about 3,800 coming down annually through at least 2022, according to Bloomberg NEF. It's going to get worse: Most were built more than a decade ago, when installations were less than a fifth of what they are now."[24]

We need to be sure we are building power systems that are reliable and will last. The power produced by wind turbines is not steady or reliable, even with all the expensive power stabilizing equipment. Locating wind turbines offshore where the winds are more reliable sounds good, but what happens when a tropical storm or hurricane sweeps through? Many experts predict worse weather events due to the increased level of CO_2 in the air. While The Get Real Program should stabilize and start to reduce CO_2 levels, storms will still develop as they always have done.

A lot of wind turbine building has been done to meet regulators' demands that utilities have a rising percentage of their power generation capability in renewable sources, regardless of whether they are producing anywhere near the rated capacity of the turbines. In many cases, a utility gets credit for the manufacturer's rated capacity in these calculations, not what the turbine actually produces. It looks like one reason wind turbines are being retired at

---

[24] Martin, Chris, ***Wind Turbine Blades Can't Be Recycled, So They're Piling Up in Landfills***, Bloomberg Green, February 5, 2020.

10 years or are producing less power is that federal tax credits run out in 10 years, making them less profitable to maintain.

There is a large incentive for companies to make products that need to be replaced often instead of selling something that lasts for 50 years or more. Wind turbine companies are no different. Electric utilities have a long-distance electric grid mindset. They focus on big, remote power generation, accepting substantial losses from the transformers and electric lines. In the case of fossil fuel, much of the energy in the fuel is wasted as waste heat. Electric utilities' plant locations and long distribution lines are not the way fossil fuel power should be produced and distributed if we are interested in using it as wisely and efficiently as possible.

- *Wind Energy Issues*

Wind energy is an area in which focusing on a single variable source to make grid-ready electricity is a problem. Wind turbine size has gone up dramatically over time because taller wind turbines capture higher velocity air that is above the Earth. Wind speed starts out near zero at the Earth's surface and on the average rises with height up to the jet stream. However, one of the problems with a very large wind turbine blade is that the tip speed is so high at even low rotational speed that the blades are deadly to flying creatures.

The wind industry minimizes bird and bat deaths saying that house cats kill far more birds; however, the relatively rare birds that the tall wind turbines kill have few young and are being sharply reduced already by wind turbines. Large-scale adoption of wind turbines could wipe out some species. Additionally, there are lots of stresses due to the higher wind speed at the top of the arc versus the bottom. Wind turbines are never good neighbors as they are quite noisy at times. A horizontal-axis turbine, the most common design, has to rotate to stay aligned with the wind axis and has to be stabilized to not shake in the wind.

Massive foundations are needed to support the mega turbines that require a large amount of both energy and CO_2 emissions to make all of the concrete. Supposedly, modern wind turbines are capable of withstanding very high winds, but we have not read of any wind farm being hit by very high winds and certainly not a tornado. Consider the time to rebuild and restore the power after a tornado or a large tornado outbreak in a core wind-power area, not including the costs and transportation time! Currently, wind is a low percentage of the grid supply and is stabilized by lots of turbine generators that produce true sine wave electricity. Wind is fickle and unpredictable. Wind does not follow laminar flow patterns but is full of eddies and turbulence, especially close to the ground. What if there were a better way to use wind to make power?

So much development is focused on making bigger and bigger horizontal-axis wind turbines. Wind turbines that are not that old are being scrapped to put up new, much larger turbines that seem to have better economics due to higher wind speeds at altitude. Old horizontal-axis wind turbines are also proving to cost more to maintain as they get older, making them unprofitable as well. Sometimes, wind turbines are scrapped when the tax credits run out after 10 years. New wind turbines are stretching the limits of bearings and materials as well as construction. While engineers can estimate life for these structures, the real test is time under the wide range of possible conditions.

There is a type of wind turbine that doesn't kill birds, and that is the vertical-axis wind turbine in which the outer edge of the blades travels only at the wind speed instead of superfast, as big horizontal wind turbines do. Sadly, the utility mindset of massive high-output wind farms works against vertical-axis wind turbines that are smaller and more suited to distributed use such as on buildings.

If we really recognized the danger of a total loss of the long-distance electrical grid, we would move away from the existing power model towards the safer distributed-generation model. Small vertical-axis wind turbines are safe for birds and bats and are better neighbors with respect to noise and height. In summary, a combination of vertical-axis wind turbines backed by methane-fired generators makes a lot of sense and engineers should recommend this option to planners.

- A Different Way to Use Wind

Using vertical-axis wind turbines means the power conversion equipment is on the ground rather than high in the air on a nacelle that rotates so that a horizontal-axis turbine's blades are perpendicular to the wind. A vertical-axis wind turbine can handle shifting wind easily. As stated before, there is a strong bias among designers to seek as large a turbine as possible, pushing the limits of materials. The longer the blades of a horizontal-axis wind turbine are, the faster the tip moves, creating more danger for flying animals.

What if we used the wind in a different way? What if we used it not as a standalone power source for fluctuating electric power, but as part of an integrated gas turbine power system that produced stable power in all wind conditions? A gas turbine works by compressing air to a high pressure and then heating it to expand its volume before it is run through an expansion power-producing stage. Typically, a standard gas turbine uses about half the power produced by the expansion stage to power the air compressor.

Now, wind power varies by the power of the wind speed, which means that there is a wide variation in power with very high-power output in higher winds. Imagine a new design gas turbine where a series of vertical-axis wind turbines compressed the air with extra fuel-powered compressors available in times of little wind. In periods of high wind speed the heating stage could be stopped completely as there would be enough wind-compressed air to fully supply the expansion turbine that is driving a power alternator to make high-quality power.

In times of average wind, the compressed air can be heated, making much more power than the wind power by itself, as a gas turbine makes twice as much power as the compression stage. This system combines wind power with heat supplied by natural gas and/or solar thermal for a steadier full-time electric generation. Having fuel-powered compressors available on standby lets the system make power in times of inadequate or no wind.

You can learn more about this system and this topic at www.fullofideas.com or the blog at www.getrealalliance.org. For the safety of birds and other flying creatures, we need to build fewer horizontal wind turbines that don't produce reliable power. Let's save our birds! Now, let's ride the waves of energy.

Wave Power

There is a huge amount of energy in ocean waves. In fact, there is more than enough to power coastal regions. Wave energy is concentrated wind energy added to the ocean over miles before it dissipates in shallow water due to friction with the ocean floor and then finally crashes into the coast. At sea, the primary energy in the wave is kinetic energy moving across the surface of the ocean with the speed being highest above the surface and gradually decreasing with depth until the deeper water is still.

Much work has been done to try and capture some of the large energy in waves, but no one has really developed and commercialized a large-scale and cost-effective way to do so. The obvious thing about waves is that they make the surface of the water irregular with peaks and valleys. Many efforts try to use the periodic up and down nature of the wave to produce power, but none so far have been widely adopted.

Trying to capture the vertical differences requires that you not flatten the wave, meaning that you only recover a tiny part of the energy in the wave. The Founder's invention company, FOI Group, LLC, has a new way to capture much of the energy in ocean waves.

The invention actually slows and flattens the waves over a distance, which can create substantial power.

Enormous kinetic energy is dissipated by waves crashing into an obstacle, but no power is produced because the force produces no movement if the obstacle is fixed. If the slowing surface is moving at less than the speed of the wave, then force is applied to it as it moves with the direction of the wave either speeding it up , or if there is a power-generating mechanism, producing power while keeping the slowing surface at a lower rate of acceleration.

The highest speed portion of the wave is the top, so it would be ideal to have the slowing surface only engage this part of the wave by being suspended by an overhead conveying system that transfers the imparted force to a power-generating mechanism. Waves in deep water have a more uniform speed profile, which is why they are more gradual. It is as waves interact with the ocean floor that they rise in height, and the top overtakes the bottom, making the lovely profile that surfers love.

Deep ocean waves actually have more energy because the energy within them is not being dissipated on the ocean floor. FOI's wave harvesting invention may be the breakthrough that is needed to capture enough energy to power the coastal areas. Go to the www.fullofideas.com website and/or www.getrealalliance.org for more information on the development of this concept. This early-stage idea is mentioned because it is an illustration of how new ideas are needed to produce long-term sustainable power rather than some ideas such as solar or wind that produce power for only a fraction of the day.

Biogas/Biofuel and Cogeneration

There is little mention of biogas and biofuel from algae in proposals to use more renewable energy, but these all provide reliable, full-time power! We do need to use more renewables, as fossil fuel is finite, but we need to be realistic about it. Some people say that we can be all electric with renewable energy. We can, in theory, but in the long term, biogas (natural gas) from waste and renewable materials derived from photosynthesis is ideal for many purposes including generating electricity. For now, we can reduce our carbon emissions by being very efficient with geologic natural gas.

Very little is said in energy literature and media about biofuels and biogas in particular. Note that natural gas is the easiest compound to make from almost any source of carbohydrates.

Cogeneration is using fuel to power an electrical power source and using the waste heat from the fuel combustion to do other useful things. For example, engineers realized that hot exhaust from a gas turbine could be used to heat a steam boiler and power a steam turbine to make more power. Doing this boosts the efficiency to over 60 percent from the base of 35 percent. This is clearly an improvement, but there is still nearly 40 percent that is lost or wasted energy.

Building big power plants out in the country means that there is no simple use for all the heat produced. It is usually just dumped into the air. It is obvious that there is a use for heat in the winter to heat buildings. One can easily see that putting a small gas power plant next to a building would make it easy to use the waste heat to heat the building, which can increase the efficiency to near 90+ percent. This could be accomplished using biogas as the gas source!

To reduce carbon emissions and make room for intermittent renewables such as solar, we need to shut down coal and build natural gas power plants, including biogas fueled designs, to provide rapidly responding backup power for wind and solar photovoltaic.

- Pipeline or No Pipeline?

To use large volumes of biologically produced natural gas in the future requires pipelines to move the gas from the biodigesters to a use point. Today, protesters often try to block natural gas pipelines; however, such pipelines would lower carbon emissions and make possible the use of biogas in the future. You have likely heard a lot about solar photovoltaic and wind but very little about biogas, even though it is reliable and available 24/7, regardless of weather.

- *Biogas Digester*

A biogas digester can be built to last many decades. Unlike a wind turbine, it works all the time and is made of simple, durable tanks – not exotic composite materials. The methane-producing bacteria in a digester are amazing, existing in large quantities in many places. Somehow, they come to exist in buried garbage, producing large quantities of methane.

Overall, we need to be much more efficient with our waste, separating it fully into different types and doing much more recycling. We have been very wasteful and inefficient in many areas. To really use renewable energy, we need to make biodigestion a big part of the mix. Almost anything containing biologic carbon compounds can be digested to make methane with anaerobic bacteria. We need biodigesters to be on every farm, converting waste biomass and animal waste into energy-rich methane for distribution to the rest of the country through our existing, and an expanded, pipeline system.

Many smaller digesters will be too small to justify a pipeline to a farm, so the biogas will have to be trucked as compressed natural gas to the nearest pipeline collection point – if it is not used as fuel for vehicles and tractors in the area.

We hear so much about the problems with CO_2 but often fail to remember that it is essential for life and photosynthesis. Using CO_2 produced by biodigestion to enhance plant growth or algae growth on a farm is a win-win. Biogas production is increasing due to favorable higher prices given by some areas for biogas as a renewable energy source. This needs to continue, along with a massive biodigester building program across America.

- Biogas for Fuel and Capturing the CO2

If we are going to have countless biodigesters making many billions of cubic feet (BCF) of biogas, that means we will have concentrated sources of CO_2 as well. This is because biodigesters produce CO_2 as well as methane. The CO_2 has to be removed from the biogas to make pure methane that can go into a pipeline or be used in a vehicle.

The methane can then be put into a pipeline or further compressed by wind power to the higher required pressure for compressed natural gas (CNG) transport by truck or use in a vehicle as fuel. So, using wind-power to compress gases has multiple beneficial uses. See more on wind compression technology and patents at www.fullofideas.com.

Using wind more productively as part of a multi-layer energy and food system on agricultural land will result in much greater energy production for all as well as more profit for farmers and ranchers. In the winter, intercooling can heat greenhouses as well as the biodigester. Biodigesters have to be heated to about 120°F to be efficient, so there is a need for a good bit of heat. This heat can be waste heat from a power plant making electricity from the methane generated at the farm or ranch.

Given a choice between building more long-distance power lines or more natural gas pipelines, natural gas pipelines make more sense. Moving the biogas to a use point that needs both electricity and low-grade heat is much more efficient due to avoiding the heat losses in electric lines. Biodigestion, using city organic waste and effluent, produces nearly as much CO_2 as natural gas (CH4). However, biodigestion is carbon-neutral since the original inputs to the waste stream started as plant matter with carbon taken from the atmosphere via photosynthesis!

To make enough biogas to significantly impact America's needs, we will have to gather basically all cattle manure, most weeds and crop waste, and grow cover crops for both food and biofuel creation. There is a lot of negativity about confined-animal feeding operations, but they are the easiest way to gather manure for digestion instead of letting it oxidize in the field. It may seem strange to city dwellers, but animals actually like being in close proximity to one another. This is because of primal instincts for defense against pack predators.

Where Get Real Alliance differs from Allan Savory is that we think that making biogas needs to be a priority and that most but not all cattle manure should be gathered and biodigested instead of being left on the ground. FOI Group, LLC, is working on inventions to gather manure and weeds from properly grazed pastures. These materials are used to make biogas with the effluent from the biodigester being used as fertilizer on the pasture.

Geothermal Energy

Our accounting and valuation systems keep us from developing one of the largest sources of energy: geothermal heat and power. Unlike the sun or the wind, geothermal heat is constant. Geothermal heat exists everywhere on the Earth at varying depths depending on geology. Certain areas, like Yellowstone Park in Wyoming and parts of Iceland, are very hot at the Earth's surface; but in most of the world there are layers of sedimentary rock from ancient oceans, which can be miles thick, standing between the surface of the Earth and the geothermal heat in and below the lithosphere (Earth's "crust").

This rock is a poor conductor of heat and is relatively cool at great depths. In other areas, there are large masses of hot granite that is difficult to drill and engineer given current methods. There has been some geothermal development and there is a good bit of geothermal power being generated in hot spots, but there is room for much more.

Geothermal energy lasts a much longer time as far as the wellbores are concerned, and its power is always available – rain or shine! The problem with using geothermal for power generation is that it is unsuitable for the existing utility model of large-scale, remote power plants. Geothermal heat is typically at a lower temperature than natural gas flame temperature, so the efficiency of a powerplant is lower, but this is overcome if there is a use for the waste heat such as heating or heat-powered air conditioning.

Geothermal heat sinks (cooling systems used in summertime) do *not* contribute to the heat-island problems many cities face on hot days as conventional a/c units actually raise the air temperatures all around our buildings and create a heat-feedback-loop! Additionally, geothermal heat pumps are more efficient than standard heat pumps in winter, especially on cold days. There need to be strong incentives for the installation of geothermal heat pumps to lower electricity usage and reduce the carbon emissions.

There can be confusion when we say geothermal. The deep heat in the Earth is somewhat well known for making power (like at the Geysers Power Station in California). Using near-surface heat sinks to provide cooling or as a heat source, however, is less well-known. Both are important and need to be used much more.

This huge energy source is the heat coming from the interior of the Earth. It exceeds the energy of all the fossil fuel reserves by more than 50,000 times. Geothermal power is 24/7 reliable and doesn't wear out. A deep wellbore costs a lot to drill but can last 100 years if properly constructed. Geothermal energy doesn't get nearly the amount of attention paid to solar photovoltaic and wind. Yet it is really the more realistic power source. There are a number of companies selling power plants that run on geothermal heat, but a small company may have a better energy system to convert the heat to power as referenced below.

Biogas can produce a lot of energy, but only a fraction of what geothermal can. The temperature of the center of the Earth rivals that of the surface of the sun. Temperatures sufficient for power production are found everywhere at a deep enough depth; however, we need improved methods of making deep boreholes that will last indefinitely to provide a nearly eternal source of useful energy for us. Tremendous amounts of lasting heat lie under the northeastern United States.

There have been experiments to try and create fractures in igneous rocks with the hydraulic fracturing methods that work so well in sedimentary rocks, which are soft and can compress under pressure. But on rigid igneous rocks, the methods haven't worked and have even caused earthquakes. As an inventor, the Founder thinks about such problems and seek solutions, some of which are mentioned in this book.

A system of boring into the granite and at a certain depth robotically creating a horizontal tunnel to another vertical shaft could work to extract heat from the granite. It is certainly a daunting task with many challenges but one that offers the hope of an eternal energy supply; it is worthy of research dollars. Another possibility is to create large chambers in the granite that can be filled with water to extract the heat from the radioactive decay in the granite.

America is a country transitioning from having declining fossil fuel resources to having a glut of oil and gas resources. This excess supply is now being exported and has driven energy prices down dramatically compared to the dire predictions of 20 years ago about $250+ per barrel oil. The low prices have lessened the concerns about developing truly long-lasting, carbon-free power that can replace the reliability and efficiencies of fossil fuel. Long-term, building housing in areas that do not have arable land but are close to geothermal resources can be super energy-efficient by not requiring long-distance transport of all essentials, including electricity!

There are certain areas with high heat levels that are close to the surface, and geothermal needs to be developed, especially in northern cities where the heat can be used during the long winter heating season and for making power. Sadly, the geothermal arena has had little of the enthusiasm and research funding given to other, less reliable renewable energy sources. One little company's creative founding genius died two years ago, but work continues on his breakthrough system for converting low-grade heat into power. See http://www.energy-concepts.com for more information.

A long-term priority focus needs to be on continuous power instead of fickle power that has to be backed up by natural gas. One of the problems with geothermal is that it is generally not a concentrated source of power, an attribute that doesn't appeal to utilities. "Local" geothermal is not going to generate hundreds of megawatts of power from one location. Also, to be really safe from a solar megastorm, we need distributed generation close to use points. Local geothermal does just that.

Local geothermal wells could be drilled in cities. In the case of places like New York City where there is very little room, wells could be drilled under the water. Geothermal power plants give off no fumes and could easily be located slightly off the land. The cold water would be a great heat sink for the waste heat from the power plant. The resulting, slightly cooled hot water coming out of the hot-side heat exchanger for the power plant can be used for building heat and maybe even absorption-air-conditioning after it leaves the power plant hot-side heat exchanger.

One of the problems with some geothermal projects is that they are designed for producing power, and thus, they waste most of the heat. Some countries, like Iceland, have hot geothermal nearly at the surface. They are able to power their economy and provide heating for the cold climate from geothermal sources.

There are two main types of geothermal heat: wet formations that may have molten rock below and hot dry rocks. Hot dry rocks, which are under most of the Earth's surface, are much more common but have low thermal conductivity. This means that one wellbore extracts only a little heat. Granite monoliths exist in many places, and the interior gets hot due to radioactive decay in the granite.

The problem with drilling into granite is that granite is very hard, and it is hard to expose enough rock face to get much heat out. Granite is so dense that even gas can't flow through it in its natural state. Large masses of granite underlie many eastern states that have huge amounts of thermal energy, but, so far, no one has developed a way to harvest that energy for use. Man is often short-sighted and often looks only at a quick return, even if the asset goes away like an oil well or a solar photovoltaic panel. We should, instead, look towards the joy of having an asset that lasts for centuries. As mentioned, a big problem is that no value is placed on fossil fuel beyond the return on the cost of extraction, so we undervalue it dramatically. Thus, electricity is very cheap. If fossil fuel were valued with the "cost of creation" included, then expensive projects such as deep geothermal and rural local geothermal projects would make more economic sense.

It is clear that using less energy is generally not a priority for people if it is inconvenient or requires a change in habits or a new purchase. This is especially the case if using less energy entails a significant upfront investment, even if the investment pays off over the long term. Ironically, standard air conditioners usually actually raise the temperature around a building and can do so even overall in a city. A standard air conditioner pumps an amount of heat outside the building greater than the amount of cool air it provides in the building. This is due to all the work done by the air conditioning compressor.

Imagine a subdivision or area of a built-up city where a geothermal well provided heat to make electricity to both heat and cool the buildings all the time, nearly forever. These small-scale distributed power systems break the utility model and might cost more upfront than what is done now with utility power, but offer zero-carbon-based, reliable, very long-term energy. Our problem as a society is that we give very little value to extremely long-lived systems if they cost much more upfront than a short-term system. In addition, most companies don't want to sell a product that lasts a very long time as it means no repeat sales.

There is a need for more research on geothermal development by government and industry, as well as preferential pricing for geothermal power. Government and state utility commissions should support more efforts in this area. Advances in horizontal drilling of shale wells offer the potential to make a geothermal well circuit. In such a circuit, a well is drilled down and then horizontally for a distance. Then, the horizontal is intercepted by a new vertical bore. Drilling in granite is very hard compared to the easy drilling done in sedimentary rocks. Intercepting a small borehole is a real trick, but the results would be outstanding – a source of heat and power that would effectively last for centuries. Eventually, the radioactive decay will deplete, but that could be in 1,000 or more years. You can learn more at https://www.thinkgeoenergy.com/.

Hydrogen

Hydrogen is a great way to capture surplus energy from wind or solar photovoltaic for use later but has drawbacks when talking about widespread standalone use. Hydrogen is the lightest element, which means it is very small and low in density. It can only realistically be stored as a high-pressure gas with a low energy density per unit volume. Hydrogen burns with an invisible flame, which makes a hydrogen fire potentially dangerous.

There are some who advocate for large-scale use of hydrogen for transport, electric power, and even air travel, but we see lots of limitations that make it more sensible to mix hydrogen into natural gas to lower the mixture's carbon emissions per unit of energy. Burning hydrogen results in a very hot flame that produces oxides of nitrogen (a pollutant) and greenhouse gas. By mixing it into methane/natural gas, the flame temperature is lower, so less pollution is created. Also, the extensive natural gas pipeline network can be used for transportation. Even if the electricity for electrolysis to make hydrogen from water is surplus and going to waste, the cost of electrolysis makes the hydrogen expensive. Pure water is needed for the process, which imposes more costs. Currently, it is a lot cheaper to make hydrogen from natural gas, which is often done at petroleum refining sites.

A so-called hydrogen economy has many limitations and would be expensive and somewhat unsafe due to the danger of handling high-pressure hydrogen. We think more research is needed and believe some serious effort is underway, but political pressures may create some adoption in the desire to decarbonize energy.

Algae

Algae production is another viable renewable energy source that is often overlooked. It needs to be a big part of farming activity including as a food source, as an energy source, and for soil enrichment. Algae are amazing converters of sunlight to useful carbohydrates, oil, and other valuable things. Using algae to consume the CO_2 produced by biodigestion is a useful step in the carbon cycle that reduces atmospheric levels.

If CO_2 is used to provide cooling and power for a building, algae tanks on the roof can provide a "living roof" to lower a building's energy needs while producing power from the algae. We will need a lot of algae to produce vehicle fuel to replace oil. Also, realistically, adjacent algae fields are needed around things like small cattle slaughterhouses as part of an integrated food- and energy-producing prairie grass system to maximize efficiencies.

The production of fat, which can be made into biofuels, has increased substantially with algae-breeding techniques based on genetic engineering. But these systems take up a lot of space. Even so, a yield of five grams of fat per square meter of tank area per day is much higher than corn's fat yield. With CO_2 fertilizing, the algae yield can increase substantially. That means about 1.5 kilograms a year of biofuel feedstock to make about a half-gallon of biodiesel.

So, a 1,000-square-meter rooftop tank could make 500 gallons of biodiesel while sharply lowering a building's heat load in the summer. There could also be other components to add value. These tests were done without CO_2 supplementation, which can double or triple production rates in large-scale algae operations where a large volume of CO_2 would be available.

For example, if integrated grass-fed cattle feeding areas were combined with a local meat processing plant, all the CO_2 created by biodigesting the cattle manure could be sent to the meat plant to be used both for refrigeration and a large-scale algae production area. A hectare of algae with CO_2 fertilization could make many times more biofuel than a hectare of corn crop, even up to 20 times more! Algae makes fats that can be made into biofuel to power vehicles and planes, but there are a lot of other things produced that can be used as feedstock for a biodigester or in some cases as a food source. Biodigesting the remaining parts of algae will produce methane and more CO_2 as well as a nitrogen-rich effluent that can be used to fertilize the algae tanks or soil for farming.

Helping plants and algae grow faster is a good end-use for CO_2, but as discussed earlier, it can be used for power generation as well. We need to think holistically about the new generation farm that is carbon-negative and produces lots of sustainable food and energy. Using biodigestion to make natural gas out of waste sets up a great situation where lots of algae tanks can use all the co-produced CO_2 to make biofuel.

Current algae farming efforts try to do it "stand alone" just using air to provide CO_2 for the algae. The algae productivity can be increased two or three times with concentrated CO_2 fertilization from a biogas system. Better yet, after the biofuel feedstock lipids are extracted, the remainder can be run through another biodigester to produce more renewable natural gas and more CO_2.

Algae is superefficient, producing up to 10 times more biofuel per acre than crops, but it needs added CO_2 to grow well. Having algae tanks at biogas digesters or CO_2 use points turns CO_2 into fuel and other products. We need to rethink how agriculture can work efficiently to be sustainable and turn farms into more regenerative operations that are net energy producers instead of big energy consumers. In order to make the carbon cycle net-negative in the atmosphere, it is far better in the long run to make biofuels from photosynthesis instead of using geologic fuel sources.

EFFICIENT LAND TRANSPORTATION

Transportation energy use for moving people and goods on land is a major component of the total energy and carbon footprint of the US and all nations, and we can do much to reduce that energy demand by rethinking how we transport things.

FOI Group and GRA are working on innovative and efficient ideas to transform land transportation and to provide renewable-energy-based, sustainable advances in transport.

We need to electrify our railroads and power them with renewable energy. Labor relations with the railroads have been tense, and railroads are eager to reduce staffing as much as possible. Historically, railroads abandoned a lot of less-used track as truck freight took over much commerce with its fast point-to-point service. Mostly bulk products and heavy items travel by rail now. However, a re-do of and electrification of rail systems can be efficient and lower overall carbon emissions while improving the environment and road conditions for all.

Using regenerative braking on the vehicles, with the excess power on downgrades either going into a vehicle battery or back in the power line, will increase efficiency. With standalone vehicles, electrical braking can be applied but is limited by the vehicle batteries' ability to be charged. With the hybrid vehicle, acceleration and stopping can be entirely electric/regenerative with little need for friction brakes. Also, battery technologies are improving.

Typical streets and highways carry a fraction of the traffic they theoretically can because of human driver issues. Traffic jams are a big waste of time and energy. Vehicles starting and stopping repetitively waste a lot of energy that's turned into heat by the brakes. With proper controls, vehicles could run very close together, cutting air resistance and allowing high speeds with very low energy consumption.

Vehicles could separate autonomously to allow new vehicles to enter or exit a travel lane. Properly designed systems can have a very long life at a reasonable cost per mile. Ideally, the vehicles should have battery packs to be able to travel on smaller roads without overhead power feed as it is impractical to have overhead wires everywhere. The batteries could recharge on a powered roadway.

Using mostly renewable electricity is ideal to appreciably lower the carbon footprint of transportation. Such a system works for both passengers and freight.

Autonomous vehicles work much better on a controlled path, and it could be that most long-distance travel is done without a driver. Passenger vehicles could be much roomier for long-distance travel. They could also have much higher energy efficiency due to the drafting effect cutting air resistance. The combination of much lower power use and the ability to power vehicles with renewable energy through electrical wires offers a huge reduction in carbon emissions! Also, the system will be ideal for moving the large quantities of rock dust from mines to remineralize the land. You can see updates on this invention at www.fullofideas.com in the future.

MORE EFFICIENT AND CLEANER OVERSEAS TRANSPORT

Transportation is a big use of energy, and man's infatuation with speed drastically increases energy consumption. Going to wind-powered sea vessels will slow transit times but will also cut fuel consumption. As was discussed, there will be a need for lots of ships to transport rock dust to areas that don't have basalt rock deposits handy. These ships should be propelled by wind and wave energy. New ship designs are needed to optimize the use of wind, solar, and wave energy.

In current ships, much wave energy is dissipated into just rocking the ship around and often actually results in ships using more fuel as the ship pushes through the waves. Currently, a lot of fuel is burned to transport people and freight across oceans by jet plane. Jet planes create a lot of pollution in the form of tiny soot particles that are dispersed in the upper atmosphere. Jet planes should be powered by CNG instead of jet fuel. The microscopic soot that is in jet engine exhaust today is so diluted by all the excess air being thrust out of the engine to propel the plane that it is invisible, but it is the major reason glaciers are melting fast as the heat-absorbing soot gets trapped in snow. A natural gas-fired plane would have no soot, and the air would be cleaner.

THE DANGER OF RELYING TOO MUCH ON LONG-DISTANCE ELECTRICITY

Very little attention is paid to the threat of a massive coronal ejection (MCE) from the sun. In such an event, long power lines would serve as antennas to capture the large electromagnetic flux brought by the solar mega-storm. Such a storm could cover most of the Earth! In fact, we don't have an upper limit on how severe one could be as the benchmark case; note when the Carrington Event happened in 1859, there was only rudimentary telegraph service, which was seriously damaged. You can find more information about the Carrington Event and solar storms on the internet. It seems wise to prepare for the certainty of such an event happening again. We have observed large solar storms more recently, but they have been directed away from Earth, so there was no effect. A long-term loss of electricity would result from a large coronal ejection in the direction of Earth. Losing electricity across most or all of the world would be catastrophic, likely resulting in the death of a large portion of the population.

Currently, there is not a stockpile of transformers and other electric equipment, so repair could take years. How long could you go without power? Distributed microgrids are much less susceptible to a coronal mass ejection and are better for efficiency. Smaller cities located close to food supplies and with sustainable design are desirable. It is highly concerning that society has become so dependent on long-distance electric power distribution that it could not survive a long-term disruption.

Ideally, future sustainable energy solutions would protect electric grids from this possible natural phenomenon, and from the man-made electromagnetic pulse (EMP) that is caused by a high-altitude nuclear explosion. For a good depiction of what life would be like in America after a CME or EMP, read the terrifying book *One Second After* by William Forstchen. The basic message is that only non-electric things would have any use after such an event, and a return to local production is required for life. See www.emptaskforce.us for more information.[25]

THE POLITICS AND ENERGY POLICY

We live in a very unequal world with citizens of countries like the U.S. using huge amounts of energy, while people in other parts of the world live very austere lives with minimal energy consumption. Even if the low energy-consuming countries continue to live in austerity, there are not enough geologic fossil fuel resources of all

[25] Forstchen, William, ***One Second After (A John Matherson Novel***, 1), Forge Books, New York, NY; November 2009.

kinds for America to continue living at its current pace for decades ahead. there is plenty of geothermal heat to sustainably boost everyone to America's level of energy use, but it will take a big investment. Geothermal could really use innovation and research to optimize it!

The trend is for low-energy countries to aspire to greater energy use as their income increases. The shale revolution has made America an energy exporter instead of its long-time role as an energy importer, but that bonanza is not enough to lift the developing nations' energy use. It is urgent that we develop ways to use less energy over time. Whether the shale revolution is killed politically or allowed to fully develop, the resource will eventually be gone. We must move towards living on both geothermal sources and the sun's energy either directly or through photosynthesis, wave energy, and wind currents. We have decades of time, so it is important not to be rash and force change through legislation.

We can lower CO_2 levels with a very modest level of donation based on fossil fuel use. America has become unsustainable in terms of energy usage long-term, but we must be careful not to shock the system and trigger a balloon-pop effect. A snowmobile can travel on top of water if it travels at high speed on land before it hits the water's edge and can travel many miles on water as long as it keeps its speed high and route straight! But it sinks if it slows or makes a sharp turn.

Our economy is kind of like a snowmobile and with the sharp slowdown in the economy due to the coronavirus, we are seeing drastic problems. Our economy relies on the velocity of money to thrive. As long as holders of debt are willing to reinvest all their interest income and return of principal back into debt rather than being paid back and keeping cash, as well as adding outside money to buy more debt, the system survives. The problem is that much of that debt is held by pension funds to pay impending retirees and will ultimately be needed as cash instead of more debt.

If we stopped issuing new government bonds because of a balanced budget, the investment money that currently goes into new bonds could be invested elsewhere such as sustainable energy and better food systems. We need investors to invest in productive things like sensible renewable energy such as geothermal or biogas and algae instead of just government debt that goes to pay for existing programs such as electric vehicle incentives and interest. As we need to ultimately live within the geothermal heat and sun's energy, so too do we need to live without continuous debt increases. One source of savings would be a more streamlined healthcare system where people take more responsibility for their own healthcare and where use of health sharing ministries predominates instead of the inefficient health insurance systems.

The world has had a major slowdown since this book was first drafted, and it remains to be seen if, sadly, the snowmobile analogy comes true. Our current crisis shows that we need to be prepared for unexpected scenarios. A disaster exceeding the coronavirus by an order of magnitude could be a major solar storm that destroyed most long-distance power lines and transformers. Some estimates are that up to 90 percent of the population could perish without any electricity.

Going to distributed power systems such as local geothermal and biogas would help us prepare for such a solar storm. Having a population nearer its food supply, with lots of local food production, will also be helpful. We currently depend on long-distance trucking for most of the country's food. What would happen if that broke down? What would happen if the major earthquake zone in the middle of the country had another major earthquake that stopped travel across the Mississippi River?

If America were to become carbon-neutral or -negative through the program outlined in this book, there wouldn't be an urgent need to move away from oil and gas. Things like mandating use of geothermal HVAC heat pumps in new construction where possible would be beneficial. Instead, home builders look for the cheapest short-term way to equip a home even though it greatly increases the home's carbon footprint.

We need long-lived, profitable investment in sustainable energy. We also need energy prices that reflect its full intrinsic value and not what a commodity exchange focused on marginal cost says it is worth. A free market—properly managed—is a good thing, but a commodity exchange-based market with options and shorting is rigged against the producers of commodities by allowing the creation of fake supply by speculators with no requirement to actually produce the commodity. Often, the trading volume of a commodity exchange far exceeds the physical supply available due to all of the created "supply" that is non-existent.

The problem is that there are many suppliers and only a few buyers of the physical commodity, so the buyers have the advantage. This challenge is particularly acute in agriculture where there are millions of farmers and just a few grain companies. It is not surprising that the grain companies are extremely wealthy, while the farmers often hover on bankruptcy. Energy prices are depressed by the countless small energy companies that must try to grow their production to pay off debt. This is the case even if the demand exceeds the supply as it does with natural gas. In theory, production should go down if the price goes down. This is the case only if the producer can afford to cut production.

A farmer, or a deeply indebted oil and gas producer who is a tiny part of a market, needs income, and if the price goes down, they will try and sell more to gain the cash needed to survive. So, a glut becomes worse, not better, and thus, often prices decline more. The solution is to ban sales of commodities on the exchange unless there is a firm availability to physically accept committed supply. If a speculator wants to sell a commodity, he must contract with a producer for that supply instead of just putting up a small deposit with a broker.

There are a lot of good designs that can really help lower housing energy needs, but they are often ignored because energy and electricity are cheap. So many spaces rely on artificial light instead of sunlight. Aerial photos of the country show cities as brilliant, bright places due to all the outdoor lights. A city like Dallas has so much nighttime lighting that the stars are barely visible. Electric utilities have encouraged nighttime lighting as they need electrical demand to use up the power produced by power plants that must stay on full-time (e.g., coal and nuclear plants).

Steam generated by an always-on heat source must be used somehow because it can't be stored. A steam turbine doesn't take much steam to run if there is not a load on it, which means that no electricity has to be generated. Some power plants even have big electric heaters that only heat the outside air when the grid doesn't demand much power from the power plant – a big waste. Texas has so much nighttime energy from all the coal and wind turbines that nighttime electric rates are near zero for power producers to sell to the grid, and there is limitation on power purchases as well. Wind turbines can just spin in the wind generating no power if the power isn't needed. Overall, we need to get rid of the coal plants in Texas and the rest of the U.S as soon as possible!

The new methods of the shale oil revolution have given the world increased supplies of both oil and natural gas. The very large shale reserves give us time to smoothly and smartly transition to renewable sources. A ban on hydraulic fracturing would end shale production and send energy prices soaring. America would become an importer of fuel again, and the world supply would be insufficient to meet demand. Additionally, such a ban would increase coal use and prevent the closure of American coal-fired power plants. Overall, such a ban would lead to increased carbon emissions and a huge drain on the American economy, far exceeding the outlays to lower CO_2 proposed in this book.

Solutions Exist

We can stretch fossil fuel supplies dramatically with the use of renewable energy. This chapter has talked about possible innovations that could make renewable energy more economic and how it could provide reliable continuous power and energy.

Biogas needs to be a big focus. Its renewable natural gas and ability to concentrate CO_2 are big positives. It needs to be put to work transferring and storing energy. There is real value in being able to boost CO_2 use at the plant or algae levels for increased productivity and sequestration of carbon in the soil. Instead of CO_2 going into the air when plants aren't growing, and the soil being bare except for rotting vegetation, CO_2 can be concentrated by making biogas out of the plant waste.

We need to think beyond current photovoltaic panels and horizontal-axis wind turbines. These sources produce fluctuating power that is not dependable. We have to move towards improved ways of providing the top-quality power that people want. Most essentially, we need a rapid phase-out of coal burning, replaced not only with big, remote natural gas power plants but with cogenerating small natural gas turbines located next to big users of the waste heat. You can see them at www.capstoneturbine.com. We can really clean the air and sharply reduce the amount of carbon going into the air with a combination of renewable energy being used as much as possible and clean natural gas backing it up.

Many proposals are unsound from an engineering perspective and disastrous from a financial perspective. During the writing of this book, a sudden crisis has emerged that really impacts America's ability to fund major change. The financial markets have gone down a lot and slowly rebounded, and many people have lost their jobs. A pandemic shows that we need to be prepared for the unexpected and plan to be more resilient. Some climate activists propose super expensive, rapid remakes of America's energy system that could be done only by government edict and would result in a very unreliable power system.

It would be ideal for the current coal-fired power plants to be replaced as soon as possible to quickly lower the carbon footprint of America and end the source of environmentally damaging pollution and the creation of so much toxic waste.

We need long-lived power generation created by long-lived equipment. It would pay dividends for the future. Society lives on the profit margin of activities. Things with large and lasting profit margins provide money for jobs and investment. Investing in things that may take years to pay back and don't last long are bad investments for investors and for America. There are many who envision a zero-carbon power system, but they generally don't put much credence into the two technologies that can match the long-lasting continuous power at an affordable price offered by fossil fuels. The two technologies are biofuel and geothermal. Clearly overall, geothermal energy needs much more attention and money put into it than is occurring now, especially for cogenerating installations in developed areas. Biofuel can benefit from greater attention and investment also in addition to focus on and development of some of the ideas included in this book.

As farmers and engineers, we are full of ideas! Together, we can develop and expand the raw ideas. To do so, we need investment for prototypes and require more people resources to bring the inventions into reality for your benefit and ours.

Carbon Sequestration Solves for the Big Issues

The Get Real Program promotes soil and plant carbon sequestration as a way to lower CO_2 levels while we continue burning fossil fuel. Pursuing these ends does not preclude making efforts to increase efficiency and reduce fossil fuel use. With the depletion of financial resources and massive debt incurred to minimize the economic damage of the coronavirus, there won't be resources for a large-scale transition to renewable power right away. It is a good thing that's not needed to really start to lower CO_2 levels. What is needed is more work by the private sector for better power solutions as well as efficiency. We need to aggressively pursue the constant sources of energy that are biogas, biofuel from algae, and geothermal. Developing biogas digesters in conjunction with algae farming for biofuel needs to be common in rural areas to use available waste for fuel.

How to pay for things is often ignored by politicians and some activists. There are proposals like the Green New Deal that call for incredibly expensive change to the way America functions without offering ideas about how to pay for it. This book promotes a "Really Green" new deal with sharply increased photosynthesis taking precedence over trying to phase out fossil fuels regardless of feasibility and cost. We have real, practical options!

THE GET REAL PROGRAM KNOWS BETTER

Many climate activists have the wrong focus. The strong anti-natural-gas movement is misguided. These activists are keeping carbon emissions high by preventing the rapid phase-out of coal-fired power generation. American carbon emissions have already gone down in large part due to switching from coal to clean natural gas. Utilizing natural gas as well as renewables is the most effective way of reducing carbon emissions.

The major long-term, full-time energy solutions are biogas, biofuel from algae, and high-temperature geothermal energy from deep in the Earth. Unlike short-lived solar farms, high-quality biodigesters and geothermal energy wells can last for a very long time and provide 24/7 energy that doesn't need backup. When we think of building new housing, it would be ideal to build it where geothermal heat is available and strongly encourage the use of the cool, near-surface region of the Earth for geothermal heat pumps. Ironically, for efficiency one could both tap the deep heat for power generation and the shallow cool region for a geothermal heat pump.

CHAPTER 10
SOLAR ENERGY

"The promotion of large-scale solar photovoltaic installations is being made by people who want to destroy America as we know it – both foreign and domestic enemies of the American way of life. We know better. It is important to advance from lessons learned in the past for a healthier and more secure future for power."

— David Munson, Jr.

How should America and the rest of the world be thoughtful in evaluating the best path forward for a sustainable energy program? How should we redirect and respond to the threats we face day after day – in the media and in the air we breathe? What is the responsible approach to our climate threats, the real ones we are facing today? Is it solar photovoltaic energy? The answer is no. Well, not *only* solar photovoltaic.

Many false witnesses around the world are advancing their own agendas by touting and reveling over solar photovoltaic's effectiveness, as if it is a natural miracle worthy of unilateral adoption to the exclusion of all other power sources. They are misleading the public on the real climate issues and other safety and sustainability concerns for America, as well as proposing ineffective approaches to implementation of solar photovoltaic programs.

The solutions we often see in the news don't solve for much and only serve to line the pockets of special interests with an endless funding need. While solar photovoltaic is a great cog in the wheel, it cannot solve for the real crisis facing our planet. There is so much more that needs to be done to create sustainability for our planet and to advance us in the climate challenges of our time. We must start with our soil, but you already know that from your study of the soil chapter, we are sure.

It is simple: we need new innovations! Not only do we need new innovations in solar photovoltaic, but there are many solutions that The Get Real Program are advancing for real impact. We shouldn't be happy with the current "best way," which has so many negative effects. We cannot trade one issue for multiple new issues that will certainly come up as a result of our myopic approach to the broad challenge of sustainability of our planet.

Climate is one big issue with many solutions, including solar photovoltaic, but more innovations are needed. It has been proven on the large scale that solar photovoltaic is unreliable power with poor quality for the electric grid that will result in lots of brownouts and blackouts. We know the issues and so do our competitor countries, but we are here to share more on the solutions – both the good and the bad ones.

For instance, if the solar photovoltaic power output drops as the sun's power drops from 700 watts per square meter to 100 watts per square meter in a matter of seconds (which is what happens when a large cloud comes across the sun), you're going to have a sharp loss of power. It is that simple. This is why building huge solar farms over a couple of hundreds of acres makes the grid that much more vulnerable to a solar disruption. Meanwhile, having small solar installations on individual roofs separated by distance means that many clouds won't block them all for very long, and you won't have this power failure.

If we rely on solar photovoltaic farms as a primary energy source, we no longer have reliable electricity. You'll have to have a battery backup in your home with quite a bit of capacity just to keep all your appliances from getting turned off and the clocks set to zero. We wish this was our only concern with solar photovoltaic advancing as a major source, but there is more...

Buyers Beware

Electricity is essentially the key ingredient for American life. Possibly over 90 percent of the people in America would die with a long-term loss of electricity, as our society is entirely dependent on it. This is why foreign countries wanting to dominate America possibly consider solar photovoltaic as an opportunity for their advantage.

Our competitor countries, like China, know the best way to destroy America is to make its power system unreliable. By doing so, we can become powerless, weak, chaotic, and utterly dysfunctional with everything halted – think COVID shutdowns on steroids when you have electricity as the key factor in the shutdown. It is crucial that we do something about this uncoordinated growth in solar photovoltaic... and now! China is watching for any opportunity to break America down, and this could be it. It could be a significant threat to America and our sovereignty that we are in some cases legislating and creating mandates around climate change that include a one-sided and limited approach, like solar photovoltaic.

Transparency is non-existent in this space. We are just supposed to "trust" that the outcomes and savings are there for us by investing more and more in solar photovoltaic. We are doing this while also depleting our backup resources in a way that will make it impossible to have easy availability if solar photovoltaic fails. The reliable infrastructure will be gone unless we determine to keep them operational.

This is what is happening as we close down energy plants and put key people out of work with each and every new regulation or executive order on climate change initiatives that focus on solar photovoltaic over traditional power sources. We become more fragile and easier to destroy without any power backup to supplement on the cloudy days or in wintertime. Also it could be a rogue nation or terrorist nation or group that, once the U.S. is overly dependent on solar photovoltaic power, launches or smuggles nuclear weapons to key points where huge ash clouds—from semi-active volcanoes that are targeted or from other geologic points that create ash and particles in the atmosphere—cover many parts of the U.S., and "nuclear winter" conditions are triggered by untraceable nuclear explosions that render the U.S. power grid useless due to a loss of the sunlight required for photovoltaic power generation. By our approach, we could be handing over our future to our competitive nations. Is that what you thought and want?

Have you noticed that the solar photovoltaic industry refuses to show charts of actual solar photovoltaic output? No one wants you to see that the new power sources that are being implemented produce half as much power in the winter as in the summer, not to mention being filled with unreliable instantaneous failures of power. What else don't we know? AND, guess who is solving America's climate crisis? It is China and Russia… China being one of the greatest perpetrators of the climate crisis? Are they making the changes that help America implement a better future for us? No. Is it time to think deeply about the advantages and progress we need here in the U.S. to keep our economy growing safely and steadily? What are our best key resource advantages?

Conventional photovoltaic farms are an environmental and economic disaster. They produce a glut of intermittent unreliable power and deface much land that is needed for greater uses. There are solutions and news ways to look at the problems that we have outlined throughout this book. We are for solar energy, but only as part of the formula solution with innovative approaches that FOI Group, LLC is advancing for real impact and change for the right problems and sustainability.

These new ways of looking at the problems from the lens of experienced farmers and engineers have created a flow of new innovations, innovations that you can read about when you visit www.fullofideas.com or subscribe to the newsletter on https://getrealalliance.org. The Get Real Program is here to educate the public in areas that no one else wants revealed. You will find your journey with us to be transparent, innovative, and full of content that will continue to grow!

Let's talk more solar...

SOLAR BASED ENERGY PRODUCTION

Utilities are particularly interested in large, concentrated power plants. If investors can get a high price for renewable power when it is generated, they are happy to invest in large solar farms even if the farm only makes power for six hours a day, has a very short peak-power time, and has only intermittent power production due to clouds. Investors don't care that this system makes the grid unsteady.

It takes about seven years for a big solar photovoltaic farm to pay back its cost of installation, and there is the potential to make several times more money before it is obsolete and has to be replaced. No one builds a solar farm thinking it will last a very long time or knowing how expensive it will be to build the next generation solar panel. In the past, the Founder's engineering company FOI Group, LLC, see www.fullofideas.com, developed a new solar energy invention that seemed to have real promise in this area.

The solar photovoltaic energy industry often hides the truth from the public. Many times, they don't reveal the true solar photovoltaic output when they promote solar energy. If they did, all would see how unreliable solar photovoltaic really is as an energy source. As part of the research and development on solar photovoltaic energy opportunities, solar radiation was measured for over a year by FOI Group, LLC.

The graph below shows a wintertime week of solar photovoltaic output. Note that when panel makers rate their output, they use nearly ideal 1,000 watts per square meter solar radiation to give their panel's output. This graph shows *actual* solar radiation, not the panel output that would be far lower! Measurements were taken minute by minute, which is why some of the days are so choppy due to intermittent cloud activity causing output to drop dramatically without warning. Note that peak radiation only occurs for a short time at less than 500 watts during winter.

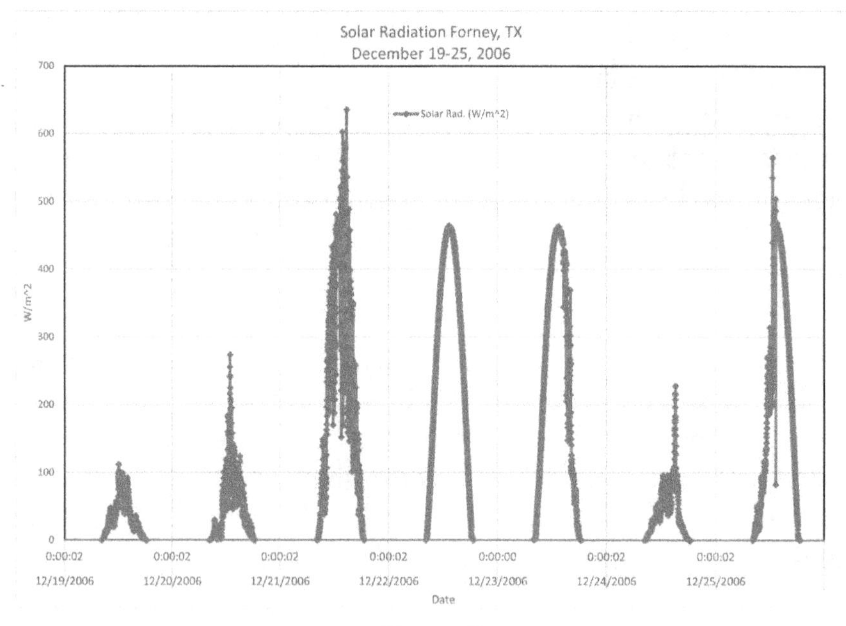

Figure 3

The ideal situation is to have solar photovoltaic power fully backed by efficient, distributed, cogenerating gas micro-turbines discussed further in this book. Many people know what rooftop solar photovoltaic looks like, but few know what cogeneration is or why we should do it. Going to rooftop solar photovoltaic combined with cogenerating micro-turbines will drastically lower carbon emissions, making it even easier to get soil sequestration to actually lower CO_2 levels. FOI's solar technology uses a more efficient solar thermal process, which totally solves the issue of intermittent clouds instantly dropping power output and allows inexpensive combustion to produce power 24 hours a day.

It is important to segregate solar photovoltaic from solar thermal, since solar thermal has built-in capacities to ride through a cloudy period. And if you have natural gas heating with backup power capacity for the solar thermal power process, it can go days without sunshine, whereas large solar photovoltaic fails so fast that large natural gas-fired power plants have to be running on standby to back it up when the power fails instantaneously. Having backup can only happen if we maintain the backup source's infrastructure and resource it appropriately for all situations.

Reliability is a key part of the interdependencies needed to keep America's grid running continuously without disruption. We have all seen in the news what disruptions can cause, and it is not good – chaos in the streets, violence, death, and other criminal activity. Power outages such as happened in Texas in the winter of 2021 caused large numbers of deaths and could have been much worse if more renewables, especially solar photovoltaic, were a large part of the grid with short-lived batteries providing so called "backup."

"The Ideal" of an Unreliable Grid

Finite Earth resources are used to make both solar and wind power systems that may be difficult to recycle. As consumers who love the reliability of electric power, we should all be very concerned about threats to a reliable grid. Having a solar energy patent and having done development work in that field, including measuring solar radiation over time, FOI Group knows how variable and undependable it is. Clouds and haze that impact the ability to generate solar power are common occurrences that can last days or weeks.

The place for solar photovoltaic panels is on land that can't produce a crop. Phasing out coal in favor of the great package of solar photovoltaic and natural gas microturbines can cut America's carbon emissions while maintaining a great quality of life.

Clouds and Solar Connection

There is speculation about what causes clouds. One theory that has some merit claims that cosmic rays from space interact with microscopic particles in the air to create them. Cosmic rays are deflected from the Earth by the solar wind and Earth's magnetic field. Their intensity is affected by where the Earth is in the solar system. Overall, we are in a time of lower solar wind and solar storms that do less to deflect cosmic rays, so cloudiness may increase. It certainly seems that there may be more clouds than when we were young. Consider reading *The Chilling Stars* by Henrik Svensmark and Nigel Calder for more interesting information on this theory.[26]

Issues with Solar Photovoltaic

There is no flywheel energy-storage effect with a solar photovoltaic farm, as there is with rotating turbine generators. The power output can't rise instantly when an extra load comes on but can go away instantly when a cloud shades the panels. This is why solar photovoltaic is not a great be-all and end-all answer for our sovereignty and the sustainability of our power grid.

Solar photovoltaic power is direct current (DC) and uses an electronic inverter to create something close to an alternating current (AC) sine wave. The difference between DC and AC is substantial. No inverter produces a totally true sine wave, but they can be pretty good. They do sufficiently well when their produced and inverted power

[26] Svensmark, Henrik and Calder, Nigel, ***The Chilling Stars: A New Theory of Climate***, Icon Books Ltd., London, 2007.

is added to a majority of power that is made up of true sine waves produced by multiple turbine-powered sources. It is like dancing – if you can't keep the beat very well but have a strong partner, they will guide you to be on the beat.

Today, an increasing amount of power used is from AC/DC converters. A converter does the opposite of an inverter, making AC into DC. Typically, these converters make low-voltage DC that is used for everything from computers to LED lights. In simple terms, they work by chopping off the high-voltage part of the sine wave, physically making a single-voltage, one-direction electron stream – amazing stuff that happens in things like your cell phone charger!

Oftentimes, designers don't think about all of the unforeseen consequences from actually changing the way things are done. How will the grid behave without the dominance of massive flywheel energy to stabilize it? And how will the new DC loads, which destabilize the sine waves by chopping off just the high-energy portion, affect the grid?

As engineers, we like the stabilizing effect of rotational inertia, which has served us so well in the electric age. It may be that having a mechanical DC motor driving an AC alternator that has flywheel capacity would stabilize the output dramatically when fast-moving clouds are sweeping across solar farms. It might cost more and be less appealing to the electrical engineers, but it may give us a more stable grid. Having solar farms put in flywheel systems to convert DC to AC will provide stabilization to the grid and allow the system to withstand a passing cloud.

Below is a graph of gross solar photovoltaic output in Dallas, Texas, during a week in July of 2007 (see Figure 4). You will notice the wide fluctuations that are shown as vertical lines all throughout the days. This is where the sun's radiation is partially shaded, resulting in rapid drops in power output. This would lead to brownouts if solar photovoltaic systems became a very big part of the electric grid.

The surge capacity of reliable, always-available turbine generators powered by natural gas or nuclear or coal cannot keep up with an ever-increasing portion of DC power input, and current solar photovoltaic farm designs have no backup.

Homeowner or small building solar photovoltaic installations are often just using the grid as backup, so installation of lots of rooftop photovoltaic will destabilize the grid during typical days such as shown on the graph. There are systems (like the Sol Ark power system) that allow a small installation to have its own batteries that can rapidly step in to keep the power flowing, but this is more expensive than the typical homeowner's solar photovoltaic system that uses the utility power as backup. A flywheel system could maintain power output through these short drops. Rapid backup power is needed from natural gas power plants, as well as for, as you can see, days with little solar photovoltaic output. Look at www.fullofideas.com for more information on solar and the different aspects of its use.

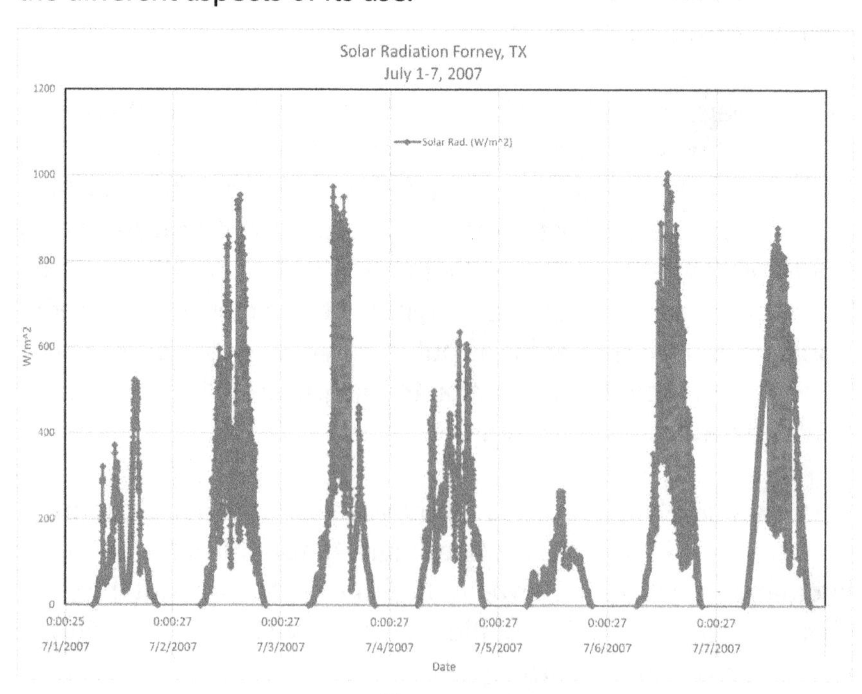

Figure 4

Behind the Scenes

Believe it or not, electronic things don't always have long lives. The inverters used at solar photovoltaic farms are considered short-life equipment that needs to be replaced every five years or so. It can take a solar photovoltaic farm seven years to make enough money to recoup the investment required to create it. Estimates are that solar farms can last 30 years, but many solar photovoltaic panels come with 20-year guarantees. Some of the early panels are still making power, so no one really knows how long the panels will last.

Solar photovoltaic panels are now mostly made overseas, including the large majority being sourced from China. This creates more dependencies for America on foreign relationships, including cost dependencies. What happens if China produces even more of our panels and sometime decides to hold back? It would take a long time to get the manufacturing capacity needed within the U.S. as a response. This would put our energy system at risk.

Little attention is paid to what must be done when panels need replacing due to age or weather damage from phenomena such as hail or windstorms. In the case of China, major carbon emissions result from making solar photovoltaic panels, due to the use of coal for much of China's power, along with the large amount of energy required for a solar photovoltaic farm. Overall, America needs to create more pathways to get what we need in order to securely run our day-to-day energy operations. The vulnerabilities became clear in the Spring of 2021 when hackers disrupted East Coast gasoline distribution and access.

More on Solar Thermal and on Photovoltaic Energy Issues

Having complete power backup for continuous power delivery should be required of anyone connected to the grid. If, as an advocate, we want to move to having a lot of solar photovoltaic without natural gas backup and short-term batteries or flywheels, the typical days shown in the above graphs (see Figs. 3 & 4) will result in very unreliable power and lack of power much of the time.

Efficiencies of concentrated solar systems (mirror arrays superheating a steam turbine) could be enhanced if combined with some natural gas superheating to make the steam turbines much more efficient. The efficiency of the process is tied to the temperature range of the process. Concentrated solar systems operate at much cooler temperatures than a natural gas-fired power plant, so they have lower efficiency. Most of the heat used by a steam turbine power plant, whether the steam is made by solar rays or fuel combustion, is used to boil the water, but the efficiency of the process is totally tied to the maximum temperature of the superheated steam.

It doesn't take much energy to raise the temperature of the steam, but you have to have a heat source hotter than the desired maximum high temperature. The vast majority of the heat needed could be provided by the solar mirrors to boil the water, then, the steam could be super-heated by a small amount of natural gas, dramatically boosting output and economics. Solar purists hate to mix fossil fuel with solar and oppose using natural gas as a backup to keep the plant making electricity at night and on cloudy days.

The good news is that renewable biogas could be used instead of geologic natural gas once enough bio-digesters are built! Solar thermal has the advantage of using steam turbines to generate the electricity, so they have that valuable rotational inertia to stabilize output and make pure sine-wave electricity. There is also a bias against the mechanical nature of solar thermal as purists really like the "no moving parts" nature of photovoltaic cells that are like black boxes with light hitting them and electricity coming out.

The problem is that it is the unconverted DC kind of electricity, and it costs a lot to convert it to AC. Also, the converted AC is not as good as a generator-created AC that has a true sine wave instead of a simulated one. It is like the difference between a vinyl record that produces perfect audio sine waves and a CD or MP3 format that produces a digital rendering of a sine wave. They are close but not the real thing.

Solar thermal has an added advantage if it is used on or near a building. For smaller installations, other types of heat engines can be used to make electricity. The waste heat from the engine can be used to heat or cool the building or both, since no heat engine is 100 percent efficient. Air conditioning is one of the biggest energy uses of most buildings. It is practically essential in warm areas, especially with modern design, so this should be a big area in which to look for efficiencies.

Mirrors are very efficient, and a high percentage of the sun's radiation gets harvested by a solar thermal system. But the percentage of efficiency drops when only electricity is the output, as the thermal engine alone has relatively low efficiency. While steam is a good solution for large systems, it requires supervision. This is affordable with a large installation but impractical for a small system on a building. A small thermal system needs to be as labor-free as a photovoltaic system, or a standard air conditioning system, to be widely adopted. FOI Group, LLC, has power solutions that work well for small installations that need development. Go to www.fullofideas.com to sign up for our newsletter and see how you can participate and support more development of reliable renewable power!

Example -

A building has 20 kilowatts (kW) of pure electric loads for lighting, computers and other small fixtures and appliances. It also needs 50 kW of air conditioning in the summer, and 35 kW of heat in the winter. The sun in this area shines brightly and gives 1,000 watts of radiation per square meter. So, the maximum load is 70 kW of electricity for a photovoltaic system with the integrated use of heat pumps. Assume 20 percent efficiency for the panels, so you get 200 watts of electricity per square meter, or 350 square meters of panels to operate the building. Now, that only provides power for about six hours a day with peak power much shorter than that. To have batteries to provide 24-hour power would mean four times the number of solar photovoltaic panels, along with expensive batteries to provide power during most of the day when the sun doesn't shine. There simply isn't enough lithium in the world to make that many batteries for universal adoption. Also, most buildings don't have enough roof area to make just a fraction of the peak daytime power needs, much less to charge batteries.

Using FOI's more efficient solar thermal system could offer a way to provide most of a building's power and air conditioning needs, as solar thermal can use the waste heat from a power cycle to operate a heat-powered air conditioning system. Solar thermal collectors can be mirrors or possibly Fresnel lenses, both of which shade a building, lowering the heat load on the building. Solar photovoltaic panels absorb solar radiation and heat the roof with heat radiating off the back of the panel.

The Backup

Another consideration is how many more collectors are needed to provide power when the sun doesn't shine. Far more collector area is needed than the basic building capacity requirement at average or even peak demand. If you need 10 kW of energy during the day on average and want to have storage to carry the building around the clock, then you may actually need more than 40 kW of solar photovoltaic panel generating capacity, assuming no losses.

It takes up to four times as big a mirror or photovoltaic array to meet 24-hour power needs, if you don't have something like natural gas to back the solar photovoltaic up. On a practical basis, there is not nearly enough lithium around to build that kind of battery backup. Most battery backup systems are just enough for part of a day and are still expensive. Companies are working on improving batteries, and work is being done on recycling lithium-ion batteries, which would make them much more sustainable. But the most efficient designs use nickel or cobalt, which are in limited supply.

Current battery storage efforts surround producing a certain amount of power for four hours, which sounds good if there is a way available to recharge the system. The same basic economic issues affect residential applications: first, adjusting for a/c timing in evenings is problematic, and second, there is no real solution to the storage/battery problem. A possible large-scale adoption of ice storage could be considered; however, it would be a major capital expense for property owners. Many buildings such as apartment complexes wouldn't have room for it as they are configured right now.

For homeowners, there is good reason to put solar photovoltaic panels on the roof with an inverter that can tie to the grid or use batteries and a generator for times when solar photovoltaic power is unavailable. Currently, batteries are much more expensive than utility power for backup, but there is peace of mind knowing that your house can function if the utility power goes down. Most homeowners choose a grid-tied system that only works if the utility grid power is on. Buying an all-in-one generator system is expensive.

Battery backup can work in a home, but the homeowner cannot use big loads on just the batteries. Having a natural gas-fueled generator is an easy way to run bigger loads when the sun doesn't shine and charge batteries on a cloudy day. As this is being written, we have just had a week of cloudy weather. Cloudy days make it obvious that solar photovoltaic can't be a total energy solution, though they can cut fossil fuel use with good planning.

ALUMINUM

FOI's patented solar energy system is truly novel and offers a new way to use solar radiation that offers high-efficiency, reliable power with heat storage and the ability to use a source of heat as backup during cloudy days or at night if demand is high. Focused solar energy can generate very high temperatures but so far no one has a workable system to convert solar energy into electricity using a thermal process on a small scale. FOI envisions using small containers of aluminum that can be moved to the focal point of a solar thermal concentrator to be heated until the aluminum melts, then moved to either storage or a use point.

The nice thing about aluminum is that it stores a lot of energy when melted and releases it when it resolidifies at a very nice temperature for efficient power generation. You can see the www.fullofideas.com website for our initial patent but several more patents are in the works, and we hope to move towards prototypes before long. This system solves the issue of intermittent outages as shown on Figs. 3 & 4 and offers a way to store energy for nighttime.

It is also easy to provide backup heat from natural gas or other heat sources for cloudy days. This system could offer a better way to use the sun for power. The good thing about aluminum as an energy storage medium is that it is abundant and inexpensive and can be cycled between solid and liquid an infinite number of times inside an enclosure as envisioned. It is also super easy to use other heat sources to melt the aluminum for those many cloudy days.

EFFICIENCIES AND SUSTAINABILITY

The two most promoted renewable energy technologies, large bird-killing wind turbines and solar photovoltaic, are not long-lived and have substantial issues with end-of-life disposal. No one discusses the need for geologic resources to make solar photovoltaic panels over and over again versus the very long possible lifetime of a resource-efficient geothermal installation. No one discusses what to do for the majority of time that the sun doesn't shine.

THE POLITICS AND ENERGY POLICY

The political push for solar photovoltaic is not the answer. You should be very concerned and suspicious of the movement to an area that is unreliable. Also, there are current examples of ineffectual energy policy: legislation recently passed in Virginia is a clear example of counterproductive, and even destructive, purely political policy. We don't want to mandate what low-carbon energy sources need to be used and in what form. The market and engineers should design the most economic and best overall solutions within America's best interests. The result of the Virginia policy will keep carbon emissions high by banning new natural gas power plants in favor of maintaining dirtier/more carbon-emitting coal plants.

Building lasting, local, intermittent power systems that can work with backup power is necessary. It makes sense to bring more solar online while we develop long-term solutions like geothermal and biogas. But there must be natural gas generation to back up the solar photovoltaic as well for when the sun doesn't shine. After taking on so much debt in the viral crisis and having consistent budget deficits, we can't afford to waste money on things that aren't reliable power producers.

Previously, natural gas prices were about half of a profitable sales price. Natural gas and oil usually exist in the same reservoir rock. If someone drills a well that produces primarily oil, it will also bring up natural gas. Perhaps 90 percent of the production income comes from the oil, so the gas is an afterthought. This produced a glut of gas with prices falling to negative numbers in some areas, while in under-supplied areas many people must use expensive and dirty heating oil. Coal-fired power plants are staying in business due to pipeline bottlenecks and active opposition by misguided environmentalists who attack natural gas power plants that could replace the coal and backup solar photovoltaic. Recently natural gas prices have recovered to sustainable levels. This should be a chance to increase supplies with responsible drilling in some of the rich U.S. natural gas fields.

As this is being written, it is a rainy day when no solar photovoltaic would be generated. What if the sun doesn't shine for several days? If solar photovoltaic were a primary source without backup, no work would be done in the dark. When people talk about 100 percent renewable energy, thinking of wind and solar photovoltaic, they are being unrealistic. Such a world is one where power is in short supply, and usage is restricted by the utilities through smart metering and by forced curtailments known as rolling brownouts and blackouts. Big users of electricity like a water heater or clothes dryer could be restricted depending on power availability. Rolling blackouts are the dramatic and destructive ways to deal with a shortage of power due to lack of sun or wind, which planners should avoid.

SOLUTIONS EXIST

Hopefully, your eyes have been opened to opportunities, while you have also come to understand that it is not practical or desirable to limit natural gas production for decades – it is essential to backing up variable and intermittent power sources, like solar photovoltaic.

Currently, there is an irrational mantra that, due to the Chinese driving down the price of solar photovoltaic panels with slave-like labor and creating lots of subsidies based on dubious claims, some other power sources (such as long-lived, always-available nuclear) are not economic. The apparently false power ratings would indicate we should buy tons of solar panels even though they don't make power when people need it – such as in the evening. Also, the heavy carbon emissions from China's dirty coal plants make solar panels not zero carbon!

We need to not be infatuated with the purported "low cost" when the solar photovoltaic panel never makes anywhere near its rated capacity in most places, as seen in the graphs earlier in this chapter. FOI Group LLC's better solar thermal system might be more expensive than solar photovoltaic panels, but it could potentially provide reliable 24/7 power with easily available supplementary natural gas heating when the sun doesn't shine for days. It could also have great energy storage ability.

THE GET REAL PROGRAM

The intermediate-term path for realistic renewable energy involves encouraging distributed natural gas microturbines and rooftop solar photovoltaic. This isn't to say that increasing the number of large, highly efficient, combined-cycle natural gas power plants is not a good idea. There are many areas in which microturbines can't be used, and it is necessary to phase out coal in order to make it easier to turn America carbon-negative and reduce other pollution.

The actions of anti-natural-gas advocates are counterproductive in the fight to lower CO_2 emissions and levels. Coal use needs to end in the U.S. and hopefully be reduced and eventually eliminated worldwide. But realistically, achieving this goal means more natural gas production as well as increased use of solar and wind. New and better ways such as FOI Group's solar power process need to be perfected. Also, what we need to do is be smart and rational and not destroy the American way of life by requiring a phaseout of oil and natural gas use. We have many decades to make a transition to non-depleting energy sources.

The Program plays an important role, but we ask for you to invest in a holistic, meaningful way. You too can experience benefits in the air you breathe and food you eat over time by contributing to real change and The Get Real Program. The Program continually strengthens as results are experienced through inventions and other experiments to transform our future. As leaders, we must all model with transparency, rather than the so-called scientists taking more and more funds without showing any real practical value. If you invested in solar photovoltaic energy, it is part of the larger picture, but more needs to be done. The "more" we need is ready to advance and could show the world true impact, but we need more people like you supporting the effort.

The Get Real Program is invested in delivering true and measurable value, not just perspectives or limited thinking on solar photovoltaic or other limited solutions. We are not against others asking for funding and running experiments in this regard, but we are in a race and cannot afford to be wasteful in how we fund efforts for improvement or to stick to solar photovoltaic efforts alone. The environment and climate are complex, but there are ways to create advantages from the complexity. This is where The Program really excels – in pulling the complex and chaotic-looking elements into real solutions that could compound into broad results!

The Program has a framework and much work completed to date, and there are so many more ideas ready to advance. The Program is a complex set of new technology, yet based on simple steps in a model that includes opportunities to evolve as we see our future vision materialize. Now is the time for solutions! Now is a good time to join this fun ride of progress for the greater joy in our tomorrows. This will take more than solar photovoltaic or even wind. We need all of our natural resources to have a holistic approach. We can do this by joining together now for "more" later. Let's turn up the HEAT on all natural energy sources in the most useful and least wasteful way!

In the end, it is important to know that we will create real impact and change when we quit building big solar farms and strictly stay with solar photovoltaic on roofs with small generators that are backed up by battery systems. We must also require that the power grid ties together with battery backups to keep a constant power output and not have catastrophic power failures. There are real solutions that are available to us. Go check out www.getrealalliance.org for more information and to invest yourself in your future and for the generations to come.

CHAPTER 11

HOW TO DO IT... PERSONAL ACCOUNTABILITY

The Get Real Program has a bold vision that is beyond what some may be able to comprehend, but every innovation starts with a vision. Vision sparks conversation. Conversation sparks action. Action sparks change. Change sparks a better, healthier, and more prosperous future for us all. While many ideas in this book appear quite bold, we can make a start towards change by spreading the thought-provoking ideas to larger audiences.

Farmers always plant the seed to bring the crops into their reality. It is time for us to start planting our seed for change!

We have to share with you that often when we hear what is happening in the world today on the news and social media, we can get disheartened. This and the closures from COVID-19 have caused us to take time for deep reflection on how we can make a difference, instead of reacting to the news of the day by spinning in anxiety, sadness, and general panic regarding our future. Like you, we sometimes feel the issues are too big for us to have an impact. That is when we know that we must do something for ourselves, the little guy and gal, and for America.

How many leaders of the past felt like the odds were against them and the issues and adversaries too big to defeat, but they charged forward out of a clear responsibility and prevailed? MANY, which is why we know that collectively we can be a great force for change and triumph. We must not surrender, but must increase the intensity of getting the right messages out with the facts to back us up. Yes, investment is required... investment in innovation, education, awareness, real solutions, and change! We must have the courage to overcome the false voices and change the perspective of the masses to create great depth and interest in order to change lives. There is much to do, and time is wasting... along with our natural resources. The job will only get bigger if we stall and wait for others to act.

Despite the rhetoric, we are ignoring real threats to our environment and planet... and all to the benefit of China and our other adversaries. They are celebrating while we are stalling progress. Think of the implications from working on the wrong solutions... wasting all of those resources that we need for real impact. We are being told to look to solutions like solar, even though anyone can see the challenges that the cloudy days will bring to the sustainability of solar photovoltaic energy as a solution. We must question the special interests and those telling us otherwise.

The answers are here in this book. America has the resources, unless legislators waste them. This must not happen. We have the information and need to make different decisions based upon the new information. We can fight for both today and tomorrow. The sacrifice of today is worth it to create a new footprint for the future.

We implore you to get involved today, as the alternative is dire! WE have to work together proactively and show our strength to combat this very real threat, not of rising CO2 levels alone... but also the loss of our fertile soils, greater risk of big wildfires, decline of the health of our oceans, and so much more. **Let's not tolerate failure; rather let each of us become a leader in the fight for bringing our planet back to its full vigor!**

THE OPTIONS.... THE WAY THAT WE SEE IT IS THAT WE ONLY HAVE TWO PATHWAYS FOR CHANGE...

Big Government or Privately-Funded Action

TAX can be an ugly word, but we have another option... personal and private actions of accountability for change! If the private sector does not take control and invest, the government will eventually have no choice; in fact, they are already driving the narrative and funding of our tax dollars towards the wrong solutions and misguided rhetoric. We can drive for real impact by our investment through private funding of the ideas laid out in this book. If we take charge now, the compounding of issues will not make it impossible to reverse and reset. Yes, it will take bold action and funding, but what is the alternative? Haven't we seen enough of big government and how it is NOT working? Can we take a chance that it works "this time," knowing the usual consequences? We wish as much as anyone that we could just sit back and wait for the planet's devastating decline to reverse magically while investing in the wrong solutions to the wrong problems. Some would love to avoid responsibility in the outcomes, think the problems are too big to impact, or just plan to push off the inevitable until later. For real, positive change, we just don't have this luxury!

Nature cannot be ignored any longer, just like we cannot sit here and hope that we don't age biologically and face declining health if we don't end bad habits and poor nutrition. We owe it to the planet to give back and make a difference. Staying at a whisper as we share our messages is not enough. We must turn the public's attention to the right things through awareness and dispute the falsity of what special interests are shoveling to us... deflecting us from the real issues and solutions! We must dispute what we know is not true and press reset on the challenge and work hard to make the world carbon-negative while we still use oil and natural gas.

We cannot let the special interests lining their pockets continue to toy with us and lead us around into the wrong fields for change… solar photovoltaic. For example, they are creating more challenges to our efforts for a healthier future through wrongful legislation, cutting the natural gas lifeline while propping up coal use in Virginia. This is a very real and major harm to our environment. We must fast-forward this country in regard to change and utilize our natural gas resources in order to have sustainability for our power system. We need all options available to us, not less. We must share our messages broadly and loudly, so that we do not see continued demise. Real leadership and education on the real facts behind the problems are key!

Think about COVID… It is so important to get treated early. It can make all of the difference in our healthy recovery and sustainability. Our planet issue with greenhouse gas is the same. The concerns are legitimate! The evidence is clear and on our side! Let's not look back, but charge forward. Let's respect our planet and our Provider. COVID should not be allowed to distract from these major long-term priorities and the urgency of beginning to act.

Instead of stopping our planet's primary areas of decline like soil, forest, and ocean health, we are worrying about eating beef and the use of our natural gas, yes NATURAL… totally wrong! We need to understand this is a private and public matter of concern and responsibility, as it impacts us all. Some we can control, but the government also controls much of the land where change is required. If we are strategic and create a wave of perseverance, we can see change in our lifetimes. We all share personal responsibility and could be on the wrong side of our history with inaction! On which side do you want to be in history?

It is all OUR collective business, not just that of the American government. We need to sympathize with the poorer nations, but also have high expectations or we all lose. We all have to work together. We must fight the urge of limited short-term thinking and not being forward-thinking, as time flies and we are losing our nutrients as we write this material. Many of the nutrients, once gone, will not make a comeback for decades or more if we do not take appropriate and scaled impact soon. Let's be dynamic and take the right actions now! Yes, we are talking big dollars, but let's do the research to learn why it will take a collective and very private effort to restore and replenish our planet through remineralization, management of our forests, and regenerative agriculture. We have to create debate to educate and promote awareness of the power of special interests that is driving the conversations not only negligently, but with damaging and dangerous consequences. They are only getting away with it because people want to believe that they are safe and that there are solutions, so in that sense any solutions are being welcomed.

The big dollars can be offered up voluntarily and conscientiously or eventually big government will have to step in… so, today we still have a choice that we may not have in our tomorrows. Companies all across America could take charge and provide charitable contributions to counterbalance their carbon impacts by offering up 15 percent of their energy usage costs. There need not be any government interference or tracking, but voluntary provision of a responsible contribution that is tax-deductible. We would prefer that the private sector act now and avoid the risk of big government getting it wrong, again. Many will not accept this at first, but public pressure can grow and make a real change in corporate behavior over time. We are already seeing the Environmental, Social and Governance ("ESG") oriented funds investing and efforts at corporations to make real ESG changes.

We all want to believe that something is being done. It is, but it is mostly the wrong solutions to the wrong problems. It is up to the private sector to cut out the corruption in our response to the threats to our planet's sustainability. It is important that we step up to share the message that the so-called solutions are only based upon who is lining their pockets and that the solutions are not creating a better world that we know could exist in our future. Let's share our message for real change!

It is the private sector, free of corruption and special interests, who can create real change. It takes just a few individuals to share this book to two handfuls of individuals and so-on who do the same in order to reach millions of decision makers with our message, so yes, we are the solution; big government is not! If we wait, the job will only get bigger, not smaller… For instance, if you do not clean up spilled milk in your car quickly, the smell just gets worse and worse and worse as it spoils… Some aspects of our Earth are like spilled milk and need a detailed fix or clean-up… a reset like that offered by The Get Real Program.

LET'S TALK REAL IMPACT

We have a big vision, including America becoming carbon-negative. Yes, that is right… carbon-NEGATIVE! It is a bold dream, but one that is possible… albeit not without you! Together, we can do it. We know the charge is quite enormous, but every innovative creation starts with a dream… and we have a dream! Let's dream together and create real impact together! Yes, we can succeed… we just have to plant the seed! Are you with us?

The Program's Bold Plan

The Get Real Program wants to drive investment into carbon sequestration through private funding... voluntary payments to pay for sequestration instead of taxing carbon usage. Again, governmental collected taxes to fund the components of The Program are an option, but the Founder does not believe that increased taxes will create as much increased impact or real change… just more of the same by big government. One proposal to cover the costs includes companies paying responsibly and voluntarily into charities that can create real change and have ready-to-go programs to reverse the impacts of our energy usage – 15 percent of the energy usage costs. Of course, we can all pay a little to gain a lot in regard to our planet's sustainability too! So, let's see how our dollars can see real impact...

Now, let's assume we want to sequester two billion tons of carbon, which is more than America emits. American fossil fuel consumption equates to 1.5 billion tons of carbon right now, and if priced at $100 per ton equates to $150 billion. There is even more from cement production and other activities. How should we spend it to get the greatest good for the investment? Here is a list of benefits:

1. **Holistic grazing and improvement of grassland**: Management changes alone can offer huge gains and increase soil carbon levels. More can be done with remineralization. We have seen tremendous changes in this area, including when soil carbon increases of 10 tons per acre per year were accomplished just with a change to regenerative grazing in Mississippi, which has virtually year-round growing conditions.

   Solution: Pay $20 per ton of proven increased carbon level.

   Potential: A billion tons of carbon or more would be sequestered.

   Cost of soil carbon testing: $100 per test on 20 acres per test for small farms and $1,000 per test for a large farm at 500 acres per test. A billion acres is about 25-30 million tests or $3.0-3.6 billion a year. Assume low added carbon sequestration per acre of one ton average.

   Program cost: About $25 billion or less for a billion tons of carbon sequestered. The carbon sequestration levels could be a lot higher, which would be great, but would increase cost. A sliding scale of pay may be necessary.

So many landowners and people in agriculture don't do soil testing. It is key to have just the right combination of nutrients for a rich and healthy soil. Testing will ensure that only what is needed is added to the soil.

2. **Pay Brazil to quit burning the Amazon rainforest and rebuild it**: Brazil currently gets no value out of leaving the rainforest alone. They don't get paid for the carbon sequestered and aren't held accountable for destroying it. They don't get charged for the large amount they emit to the atmosphere by burning and clearing either. Other countries around Brazil can be included as applicable.

   Twenty million people live a subsistence existence doing great environmental harm for very little money earned in Brazil. It is necessary to pay Brazil and its settlers to quit clearing the rainforest and an additional amount for replanting and remineralizing it. Part of the reason more rainforest is burned and cleared each year is that the land cleared quickly declines in production. Remineralizing open acres will boost productivity and cause soil to sequester more carbon. This program really needs international support to do more to restore the rainforest – by initiating a large-scale remineralization, for example.

   <u>Program cost</u>: $20 billion a year to keep a million hectares (a hectare is an area of land that is about the size of 2.5 football fields) or 250 million tons of carbon from being burned back into air and allowing 10 million tons of carbon to be sequestered each year by the saved rainforest. Ideally, this program is funded by other countries pro-rata. U.S. money would be primarily allocated to The Program overall including more rock dust and biochar in America, and the Rainforest plan would be a component funded by government tax receipts.

3. **Solar biochar-production grant programs**: Buy numerous solar biomass/biochar processing units and fund distribution increments each year at an annual cost of $10 billion. Ongoing operating payments start at a billion dollars and rise by a billion every year. Sequestration amounts would be five million tons of carbon the first year.

   Advantage: Permanent removal from the carbon cycle and increased retention of soil nutrients with less runoff into the ocean. At the time of publication, this is a conceptual invention and needs validation before being adopted as a major program; it may be found that other methods will prove to be more practical.

   Program cost: $11 billion for the first year with long-lasting value, sequesters 100 million tons over 20 years with ongoing maintenance fees.

4. **Rock dust remineralization**: Rock dust remineralization will increase photosynthesis and soil life, leading to more carbon sequestration.

   Program cost – estimated: $100 per ton of rock dust applied or more depending on freight. Estimate one extra ton of carbon sequestered per ton of rock dust per year with increased soil and plant life. Very cost-effective, as sequestration continues and trees stay healthy longer, preventing carbon from returning into the air.

   Start at 100 million tons and increase (starting on lowest cost per ton for land that is close to mines). This program is a major job and business booster, with many trucks and even a fleet of airships required to deliver rock dust to forests and inaccessible areas that would profit from the improved growth. A Rock Dust program needs to grow substantially over time to be a bigger part of the sequestration program.

First year program cost: $10 billion. This needs to increase substantially to be able to remineralize all American land.

5. **Planting cover crops**: These help in two ways. First, the growing crop sequesters carbon in its biomass. But most of that carbon returns to the atmosphere, only to be reabsorbed by the following crop. The second advantage provided by cover crops is that having living roots feeding soil microbes, and thus insects, and thus animals, prevents much of the consumption of the primary crop root mass and related carbon compounds for food. They continue building organic compounds that can permanently keep carbon in the soil.

   An estimated five tons of extra carbon and prevented soil oxidation per year per acre for 300 million acres, equates to 1.5 billion tons of carbon sequestered or not emitted to the atmosphere. Pay farmers $100 per acre for cover crops.

   Testing cost – soil: $1 billion (with some savings with simultaneous testing for holistic grazing/management tests for joint owned/managed acreage).

   Program cost: $11 billion per year initially, rising to $31 billion for up to 1.5 billion tons sequestered. With a several year ramp-up, the program would have reduced payments over time to allow more biochar and remineralization. Cover crops are a net gain for farmers but have startup costs and require change from traditional multi-generation farming methods that will take multiple seasons to fully implement. Grants or subsidies will be needed for a short time.

6. **Arid land rejuvenation in America**: Restoring fertility and modifying land to trap and store water with increased planted vegetation is a worthy cause for man's long-term survival, but its per-acre cost is expensive. Much of the American West features various desert scrub bushes that

have very slow growth rates. Hand-trimming them and making biochar will be slow, but it has an added benefit of providing biochar for enhanced water catchment to sustain more carbon-sequestering grasses and other plants.

Money needs to be expended on research and plant development. There is a question of whether to improve land with an eye towards holistic grazing and/or biofuel production as well. The use of airships and other innovations could enable large-scale rejuvenation at a lower cost per acre. One of the challenges of arid land reclamation is where to house and take care of all the people needed to work on land to improve fertility, water absorption, and planting on land that may be far from a road. Much of the arid western states are very rugged and unsuitable for improvement because there is literally no soil, and the land is steep.

<u>First year program cost</u>: $1 billion for research and trial work. More cost will be incurred in later years as other programs wind down.

7. **Carbon reduction programs**: We need to encourage people to reduce the amount that we need to sequester through their daily activities on a voluntary basis. We need to make and utilize more efficient geothermal heat pumps, better light bulbs, and more efficient cars. Electric cars are not zero carbon, because they require enormous amounts of fossil fuel to produce and have short lifespans, requiring replacement.

   More efficient hybrid cars are actually a lower carbon footprint overall than electric vehicles and provide the advantages of fossil fuel or renewable biofuel to provide long-range and quick refueling. The majority of people in America don't park their car in a garage where they can have a charger. There's a tremendous problem charging electric vehicles, including the fact that the electric grid is unable to handle the extra demand! Also, there will be major problems with trying to replace an electric fleet's batteries at their end-of-life term.

We have to also think of the impact from our changes to a more prosperous planet and existence, like that of displaced workers such as coal miners. They will need some assistance to transition to new, better jobs, and new jobs will need to be created in coal mining states. Hopefully, there will be basalt rock deposits that can be mined for rock dust that should be a growing industry that uses mining skills. There are about 50,000 coal miners still working, and 32,000 of them work underground.

A lot of American coal is exported. Even with a carbon sequestration payment of $50 per ton, this coal will be sold to countries that have no immediate option to shift from coal burning for their power. Another issue is coal miner pensions, as many miners are covered by a company pension. Factories to build biochar facility components could be placed in coal mining country to provide new jobs. Paying generous retraining benefits is probably needed, politically and ethically.

The reduction in carbon emissions by switching from coal burning to a mixture of renewables and natural gas is enormous and makes paying miners a very cost-effective part of an overall atmospheric CO_2 reduction program. Paying $100,000 per worker is $5 billion over several years to get them working sustainably. Politically—and humanely—we need to fund the coal miner pension program as well, perhaps not with a lump sum but agreeing to pay in, as needed, in the future.

Legislation effectively making coal totally uneconomic is somewhat confiscatory, like eminent domain projects in general, and as part of the program, the coal companies should be bought at a low, fair price based on current values. The mines could be saved/preserved for posterity in case there is ever a need for coal.

Program cost: $10 billion. A coal buyout is a transitional expense that is short-term for the most part. If this funding is made available through taxes on coal, the funds can also be used for remineralization in later years.

8. **Foreign arid land improvement and food programs**: Paying people in Africa and Latin America to do soil improvement and remineralization is very cost-effective and humanitarian. People can't work if they are malnourished; providing food to people in exchange for work—like making water catchments and biochar—does a lot. Fighting desertification is vital for humanity to thrive. Instituting large-scale adoption of holistic grazing will rebuild grasslands.

   Program cost: $10 billion a year to treat 100 million acres. Improve carbon sequestration by one ton per acre per year or more. This would be a program that should be funded at a much higher rate by countries with little land available for carbon sequestration. With a global program, this money would be used for more remineralization in the U.S.

9. **Ocean and marine wetland improvements**: Remineralizing the ocean and marine wetlands, as well as bringing restoration, offers great benefits on a lot of levels. The proven biorock technology really creates and stimulates reefs and other carbon-sequestering life. Large-scale remineralization with the right minerals can really increase carbon sequestration and marine growth. We need to make the ocean more productive to sustainably provide more food for a growing population.

Program cost: $2 billion to start, more after the initial first year ramp-up.

10. **Temperate but demineralized country remineralization**: Soil that is leached and has year-round growth gets very demineralized, and many countries to the south of the U.S. suffer low agricultural productivity in areas where there is not much volcanic soil. Spending money on remineralization with basalt rock dust and other amendments can sequester carbon and boost income with more food production. This economic boost will help diminish the need to emigrate.

    Program cost: Estimate at least 10 million acres at $100 per acre for $1 billion. This is merely a pilot program that is a fraction of what a global program would do.

11. **Forest land carbon sequestration**: Healthy forests are essential for carbon sequestration and rewarding tree planting is important. Estimating additional growth from remineralization is tricky, and to be effective and cost efficient, we can't pay much for tree growth... perhaps $5 per ton on private lands with a repayment clause when logged. This is fair since most forests are already in place and doing their part to sequester man's emissions. Discounts or credits for use of rock dust when replanting would also be good. Restoring U.S. federal forest land is also badly needed and will increase costs.

    Program cost: Allocate $10 billion for timber programs but more could be needed.

So, for about $150 billion a year (after a ramp-up with lower amounts to begin) we can estimate sequestration of 3 billion tons or more of carbon a year for an average cost of less than $50 per ton. Now remember, this is carbon being sequestered, not CO_2 being removed. CO_2 is two-thirds oxygen so that translates to removing twice as much CO_2 as America emits with its fossil fuel use, or nearly a third of total global fossil fuel emissions. Some measures are not permanent removal from the carbon cycle but are cheap processes per ton, and if sustained, do lower atmospheric levels until reversed – if they are ever reversed. Others, such as biochar, are more expensive but their benefits are nearly permanent.

Note, the expensive programs that don't sequester much carbon in the first year have lasting valuable benefits that can grow over time as the carbon sequestration revenue falls due to less fossil fuel use. Things like holistic grazing and cover crops take money and a management mindset change to get started, but once they're implemented, they produce substantial economic gain that will further improve rural life and profitability. This estimate is conservative but shows the incredible potential that nature has to reverse atmospheric carbon rise.

There have been numerous reports generated that offer much lower carbon sequestration numbers, but there is a widespread lack of knowledge about the benefits of remineralization or even holistic grazing. Despite decades of success restoring grasslands, Allan Savory's methods are still not widely adopted. This proposal is just a start on what a national program will do and needs a lot of input and research to optimize it.

Many of the steps to help get The Get Real Program enacted (like supporting the coal miners and compensating the coal companies for being put out of business) require government support and action. The Program wants to handle the coal miners humanely under the new way to a better future that is being proposed. All of this is intended to go away in future years and allow more to be spent on remineralization and forest reclamation. Emissions will decrease significantly with improved efficiency and elimination of coal in the U.S.

A GLOBAL PROGRAM

A global program is needed with the largest amount spent on remineralization because so much of the world is very badly demineralized, including the precious Amazon. Sharply higher timber growth around the world can play a big part in making the world carbon-negative. We can eliminate world hunger in a sustainable and carbon-negative way with a move to regenerative agriculture and holistic grazing. The world is a big place with a wide range of economic conditions, energy use, and potential for carbon sequestration.

We truly need energy-consuming, high-density countries with little fertile ground to subsidize carbon sequestration efforts in countries that have lots of land but very little money. The easiest way to collect a global carbon emissions payment is from the seller of fossil fuel or large carbon emitter such as a cement factory and distribute the money to areas of high carbon sequestration potential. Here are some solutions to rising CO2 and other more obvious problems.

A global voluntary payment program at $100 per ton of carbon would generate $1.2 trillion a year, which can really change the world for the better over some years. We can have a strongly carbon-negative atmospheric change while we still use oil and natural gas and even some coal for a while with a bold program outlined briefly here:

1. **Increasing arable land size and quality**

   We need land to be made more fertile not by clearing forests or reducing wildlife refuges, but by fighting desertification and improving the quality of the soil with soil testing and remineralization with both focused and broad-spectrum minerals. The amount of arable land in the world is very large although it is sadly decreasing by about a million acres a year due to growing deserts. Man has actually hurt soil quality through millennia of harmful agricultural practices and failure to remineralize. Entire civilizations have disappeared due to the loss of arable land. Deserts once were fewer and smaller but have grown to be more than a third of the total land mass.

Ideally, soils should have major minerals in the proper proportion along with all the trace elements. Calcium and magnesium are the essential major nutrients but sadly, deposits of these minerals are not found everywhere, so it may not be possible to bring all the world's soil into an ideal balance of nutrients.

Remineralizing the land is a huge project that will be the largest global effort undertaken by man but have so many positive benefits that make it a bargain. Because the area needing help is so large, we need to devote a large majority of carbon sequestration revenue to it. Doing soil testing on land suitable for growing crops and adding needed major nutrients along with concentrated trace elements as well as broad-spectrum basalt rock dust is the primary action we need. This will sharply boost food production in quantity and quality. We will go from being a world with hunger to one of plenty. This is a huge project requiring a major expansion of mining the needed rocks and transporting and spreading the powdered rock.

Grinding rocks to powder creates a large surface area that soil life can attach to and from which it can extract the rock minerals for other soil life and plants to use. However, most methods of grinding rocks use lots of energy and create small dust that can cloud the air, perhaps hurting people and nature. Rocks that are useful for remineralizing land are not found everywhere and are often most badly needed where they are not found. Rock dust will have to be transported from where it can be mined properly to the areas where it is needed.

New methods of transporting and grinding rocks are needed that are not major users of fossil fuel and are environmentally benign. While The Program eliminates the need to phase out fossil fuel use, there isn't any major excess supply and a massive remineralization program would take lots of energy to mine, grind, and move hundreds of millions of tons a year. Building the machinery needed would take a big chunk of the carbon dollars but would also have lasting effects from decades of remineralization covering all the arable land and the oceans.

To undertake man's greatest task will take real innovation and The Get Real Alliance has lots of good innovations discussed in this book. Your donations can help fund research and development.

A large fleet of ocean-going vessels will be needed. For example, a vessel holding 100,000 tons of mid-sized rock crushing 100 tons an hour will need 1,000 hours to crush it. If the crushing is powered by solar energy for six hours a day, that is 160 sunny days of crushing. The vessel can travel very slowly and be designed to have nice housing for its workers. In reality, the vessel may be a very different marine design that resembles a collection of floating structures rather than a traditional ship, with some parts operating autonomously.

Wave energy would allow round-the-clock grinding, sharply reducing the time needed to grind the rock. Faster rock grinding would allow a ship to grind much more rock a year, which is good because grinding 500 million tons a year would mean 500 wave-energy powered vessels rather than 2,500 solar powered ones!

As most of the world has no roads, ports, or railroads, using new design lighter-than-air craft to transport and distribute the pelletized rock dust seems like a solution.

Regarding soil testing, you can read more about comprehensive effective soil testing that we used to boost the production of the Founder's commercial grass-fed beef operations to very high levels of quantity and quality at www.kinseyag.com.[27]

Unfortunately, most soil researchers use chemical farming methodologies that often produce large quantities of deficient and sickly crops that need toxic rescue chemistry to survive. Researchers often focused on expensive proprietary products, not inexpensive rock dusts. Healthy soil produces healthy plants that resist disease and parasites. This is facilitated by the Get Real remineralization plan.

The second use of rock dust, in particular basalt rock dust, as promoted by www.remineralize.org after farming and crop enhancement is on grasslands and forest. Amazing increases in production and quality have been measured in demonstrations. With modern technology, the airships can be remotely piloted so they can operate around the clock transporting and spreading the rock dust.

On demineralized soils, growth rate increases of several hundred percent have been measured in seedling trees so the ability to fundamentally increase carbon sequestration by forests can send CO2 levels back to pre-industrial numbers even while we use fossil fuels. The unexpected problem maybe reducing CO2 levels so much that food yields are reduced, but that problem is a long way off in the future as Earth's population continues to increase. By then, space-based and other solutions may have a real impact.

As an estimate and to reflect its critical importance, using 50 percent of the carbon funds to accomplish remineralization

[27] https://www.kinseyag.com

over perhaps 15 years, with supplemental work continuing on to improve things further appears prudent. That is a billion tons a year at $600 per ton initially. While that sounds high, to be really effective the rock dust needs to be mixed into the soil and some of that work may be manual in forests.

For example, in the Brazilian rainforest, paying people to remineralize will preserve and enhance the growth. Much of the expense is front-loaded – building the needed vessels, airships, and other equipment with lower expenses in future years.

2. **Biochar**

Biochar, as discussed in the national program, is a miracle product that was discovered thousands of years ago in Latin America but largely lost until recently. Biochar is biomass—comprised of plants and trees—that has been burned in an oxygen-deprived low-temperature fire that leaves only pure carbon as char. Biochar is a sponge for nutrients and water as well as a habitat for soil life.

Biochar can be made in many ways and needs to be widely adopted globally. As a lot of the world uses wood for fuel, biochar-making stoves and heaters can be funded to create biochar as a byproduct of wood-burning stoves. Sharply higher timber growth from remineralization will allow fuel wood plantations to be planted that fight off desertification with water-efficient trees. The increased use of agroforestry will increase the amount of wood available for making biochar.

There needs to be a massive program to remove deadwood from forests to make biochar instead of letting it release its carbon back into the air by rot or fire. There is a need for lots of innovation to harvest the deadwood, make it into biochar, and move it to land that needs it. New designs of aerial timber management systems and airships for transport seem likely solutions. As a crude guess, perhaps 25 percent of The Program funding goes to biochar and timber rescue, or about $300 billion a year.

Historically, man has neglected and abused forests, - it is believed that a huge desert, the Sahara, is located where once a huge rainforest existed. We must invest in trees and the benefits will be great. A good bit of these funds can go to preserve rainforests around the world, making them a benefit for the host country instead of just unused land.

3. **Holistic grazing**

Holistic grazing is the most effective way to upgrade grassland and fight the loss to desert. It is perhaps the cheapest way to sequester a lot of carbon but takes a mind change in the way we manage grazing animals. Around the world, herdsmen aren't well paid and so engage in labor-saving practices that destroy the grassland. Providing supplemental pay for herdsmen to follow holistic grazing, which requires more people to keep the animals moving to new pasture and allowing the grazed area to recover fully before being grazed again, is needed.

The great news is that far more animals can be grazed so food supply is increased. Combining remineralization with holistic grazing makes the grasslands a carbon sponge. Estimate just five percent for holistic grazing support, mostly for herdsmen, which will really help in very poor countries. The proven methods of Allan Savory need to be taught globally.

4. **Marine improvements**

So much can be done with improvements to marine wetlands and reefs to increase carbon sequestration and marine life. We face a crisis in the overfished and abused oceans. Many acres of marine wetlands have been destroyed or degraded. Applying rock dust to them can really help boost productivity.

The proven effective biorock technology uses low-voltage DC power on a light steel lattice to sharply boost coral and marine life. Biorock even boosts marine plants. We need a massive biorock construction program on existing reefs, new artificial reefs, and some new marine wetlands. Assume five to 10 percent for boosting the marine environment to be a bigger carbon sponge and boost production.

5. **Improved agriculture**

Paying farmers globally to sequester more carbon by keeping the soil life growing and becoming more regenerative can really lower atmospheric CO2. Much is lost with the bad-but-typical agricultural practices of keeping the soil bare for most of the year and not growing a polyculture of plants. Great increases in productivity can be had while sequestering billions of tons of carbon.

We need to pay our global farmers better to preserve the soil and boost soil carbon levels dramatically. Estimate 20 percent for a huge drop in carbon levels combined with remineralization and holistic management, which includes grazing animals in more farming situations.

You will notice some similarities between the national and international program, as well as some differences. The big difference is that globally, the big emitters often don't have productive land to improve, so there needs to be a big transfer of money from high-emitters to high-potential sequestering rural countries. Both programs are just starting points that need a lot more research, development, and improvement before there is a chance of legislation passing.

Unlike the Paris Accord, we use natural forces to really change the atmosphere while also improving life. Sadly, so many just focus on demonizing fossil fuel emissions, thinking that everything will be alright if we just stop using them, but actually the really serious problems of soil disappearance and environmental degradation need to be addressed. Solving them makes it impossible for carbon levels to remain high, no matter how much oil and gas we use!

We may not be able to get to all of our issues and solutions right away within the private sector, and some solutions require the government to manage their own property responsibilities, but this book lays out what is needed. Let's all do our part to start...

START TODAY BY...

... sharing this book with 10 friends, who can each share the book with 10 more. The profits from the book will be invested into these real solutions. Getting the word out and educating those around you can drive further actions and impact.

... sharing the messages even further on social media forums to get the word on real solutions out widely.

... educating legislators, family, and friends on real solutions to the real problems.

... driving opportunities for respectful debate in public forums.

... investing in credible organizations creating real impact at www.remineralize.org and www.getrealaboutclimate.org.

... signing up for the newsletter to stay abreast of the newest information on www.getrealalliance.org.

Change is coming, whether we like it or not and whether we take action or not. Daily, we are creating more challenges by not working on the right solutions or focusing on the real issues. So, if action will be required at some point in the future, why not be proactive to avoid greater sacrifice later? If we don't start now, at some point, actions towards change will no longer be a choice. Let's avoid mandates of unneeded and unproductive sacrifice by taking control and providing leadership privately to avoid big government forcing limited freedoms, high taxes, and other personal challenges in our day-to-day lives. We, the private sector, can be the game-changer and drive the conversation today. Are you ready?

CHAPTER 12

FREEDOM IN GENERATING CHANGE FOR REAL IMPACT

What if…

> … there is a way to make the world carbon-negative…
>
> … in a sustainable way that creates little disruption for the masses…
>
> … while we still use oil and gas in targeted ways that cost less…
>
> … and importantly also solves the other very tangible problems of declining soil fertility, massive soil loss, growing deserts, ocean decline, fewer nutrient-rich foods, and declining biodiversity and dying forests?

The time has come to share the most important secret ingredient that has been hinted at throughout this book. The Get Real Program is here to share the message and create action for the climate solution. Is more BIG government or HIGH TAXES the solution? NO! Big government is never a solution, and is actually always central to the issues, as you will see is the case. We wish we could say otherwise, because a government having so much power and funding could create big changes… and fast. Unfortunately, we know that won't happen. We have case study after case study to prove it. Then, what is the solution?

You are exactly right: the solution lies within us… YOU and ME! The world needs intelligent people solving for what really matters for the sustainability of the planet, not just focusing on so-called global warming and misled conversations on climate change. Your support can be what starts the generation of change for real impact. Without you, the vision of The Get Real Alliance will remain just that and nothing more – a vision. We need a massive broad-based program of private charity by concerned individuals and companies to undertake the necessary carbon sequestration and create unstoppable momentum.

There is not a global government to do it, so The Program needs to be privately led; however, some countries may choose taxation. The challenge The Program will fight is the misguided belief that natural carbon sequestration can't do enough to make the world carbon-negative. Generous support of www.getrealaboutclimate.org and all the other charities mentioned in the book can really make a difference in helping the world get behind restoring the Earth's ecosystem to health. The Get Real Alliance, small government supporters, do not want to see a huge global bureaucracy spring up that wastes a lot of the money, while private efforts are more efficient and need a fraction of the money. We estimate that if all concerned people and companies gave about 15 percent of their energy usage costs to the right charities, we could more than offset the carbon emissions of man, making the world carbon-negative. We do need to pay more for many things to change production to a sustainable model. Cheap items made in an unsustainable way only lead to shortages in the future.

Let's Get Real About Climate:

It is our collective Responsibility to Educate ourselves and take Action to create Lasting impact on our environment!

Have you been wondering what you can do to create real change as you read this book and are inspired to make a difference in this world for yourself, your loved ones, and those yet unborn? Education is key, which is why The Get Real Program is investing time and energy in this material to demonstrate and build growth of program knowledge in the public. We all have a responsibility to educate not only ourselves, our loved ones, and our friends, but also anyone else who will listen. Sitting quietly in the corner will not affect change. Your action is needed.

You may wonder what one person can do to make an impact. Well, The Get Real Alliance started with one person, David Munson, who leads the effort to share the message. You, as one person, can buy this book for 10 close friends and family. If they in turn buy it for 10 friends and family and the multiplier of 10 continues to trend, your initial purchase could help lead to billions hearing the messages in this book! Yes, we just said that you can help impact BILLIONS of people with a small contribution, but let's not stop there... just think of the impact... as we are also using the proceeds of the book to further innovations in this area.

Taking the word of those "reporting" the information in media forums will not provide the real answers. We need to work together to get this very important and truthful information to as many people as we can. We need to get this book into the hands of the politicians, so they can see through the lies and misguided arguments of special interests lining their pockets. We need to share this book with friends, so that they can see what is happening to the food they eat and the water they drink, which is slowly impacting the planet and all of us in so many detrimental ways. It is our responsibility to act now for the sake of our children, and their children, and so on...

Let's work together to solve for the right solutions... at a minimum. One of the big problems with most climate proposals is that they only slow the rise of CO2 at a huge cost. The rising costs will be devastating to the poor, so they will likely have to cut even more corners in their lifestyle and most importantly in nutrition. Stopping the degradation of our natural resources is crucial, but many of the innovations and solutions throughout this book will serve far more purposes than just being a disrupter to the rising CO2 levels. Let's work together and make America carbon-negative!

We highlight a lot of great information in this book, as a starter for the conversations, but there is so much more to follow. On top of sharing this message through mass distribution of this book, also consider how you can advance the numbers on the mailing list for continued education of the masses at www.getrealalliance.org. It is critical that we don't keep this information to ourselves or allow social media or others to bury it. The Program has a vision that this message will be shared with the masses, but that will certainly bring out the naysayers and those with propaganda that does not serve the public. Let's share the book and other messages on Get Real About Climate so broadly that there is a public forum for debate on the solutions… real solutions! We look forward to the debate!

Those with propaganda will be "caught with their pants down," as they say, if they challenge the material in this book or the solutions contained on the website www.fullofideas.com, so let's make this happen. There are many, many so-called experts running around asking for funds to invest in what we already know does not work. We are not even sure if they know enough about the problems to have solutions or if they are purposefully misleading the universities and government funders as well as the public.

Again, LET'S START THE CALL FOR PUBLIC DEBATE! We hear the leaders on climate change TELL us what they want us to believe, but the debate of the information offered is less than impressive. Through an exchange of words, we know the so-called climate leaders of the day will reveal their "special interests" and lack of knowledge and concern over the real problems. We are READY! Aren't you?

The Get Real Program is calling for a revolution in thinking and action and is here to give you tools for continuous and consistent delivery of meaningful information to improve America's current climate situation and take control of the narrative, not just the gimmicks shared on the topic. It is important to share our knowledge in ways that are useful, memorable, repeatable, and advancing the conversation and solutions towards real impact. We know there is opportunity to join with YOU to enhance the scope of where The Program can grow and make an impact. We want to make sure the real solutions do not fall through the cracks while we are watching solar photovoltaic advance unreliably and at the benefit of foreign nations not so friendly to America.

Let's challenge the messengers of today. We know there are criticizers out there pushing down inaccurate data and information to build improper perceptions about where America needs to go to gain energy prominence and independence. Some lab researchers yell about our eating meat, yet they fly to international meetings, creating large amounts of carbon emissions! Some of the public may have innocently gained misperceptions and others have agendas that are not in America's favor. We need to share the truth and the data to back it up. Many bad actors are out there grabbing dollars and pushing out inaccurate information to line their pockets. We need to shut them down with the facts and real solutions for change!

The best time for creating real impact was generations ago, but the next best time is now. First, we need you to help lay the foundation to advance our energy systems and build from there, rather than just cutting out entire sources of energy overnight in hopes of replacing them with unreliable sources like solar photovoltaic. There is a give and take. We are all responsible for giving back, as we all work together to impact and partake in the energy solutions out there today.

WHO CAN SOLVE FOR THE BIG ISSUES?

In order to advance America's leadership in climate, please consider aligning with The Get Real Program's efforts to educate and create real impact. The goal is to solve the climate issues of today and so much more by voluntary private and company actions that make the world sequester more of the carbon that man emits, while we still use our essential resources of oil and gas in a targeted and effective way.

BIG government is not the solution! Non-profits are key to effective impact and leadership, creating a global carbon sequestration program. It is a big idea, but the best and only true solution for real change. Good things can happen with far less funding through nonprofits, rather than any government behemoth that creates the wrong solutions to the wrong problems with a heavy bureaucratic cost. Private enterprise in concert with non-profit action can have more flexibility and drive for change without the misdirection of special interests' foothold in our political leaders.

It takes "we the people," collectively, to succeed, not the few appointed government authorities with the connections and heavy lobbyists but no natural expertise! We can accomplish this together by getting the word out to the masses through this book and social media outlets! What is your conscience telling you to do? Will you help?

The Get Real Program Has a Message for the Masses
A Practical and Real Impact Approach
to So-Called "Climate Change"

The material throughout this book can get a bit heavy for those not entrenched in it day in and day out, so we thought it would be helpful to share the core messages on the problems and solutions in an easily digestible way to help the message "have wings." The material below does not have the depth of the chapters but serves as a reminder and reference for some of the fundamental messages of the book.

It is key to remember that the overarching goal of sharing this book is to get the word out that America and the rest of the world can become carbon-NEGATIVE through carbon sequestration techniques, remineralization, and other regenerative efforts. We can do this without huge impositions that are being thrown out there in public conversations today. We can phase out coal over time and maximize the use of our natural resources of oil and gas where it makes sense for energy independence and sustainability. Now let's dive back into the details…

The Big Give

Many people don't think the rising CO2 levels are really a problem worth big sacrifices—fundamentally changing our way of life—to solve. The people who do think CO2 is a big problem propose super expensive and life-changing solutions that only minimize the CO2 rise and perceived potential damages. Instead of activists trying to force big change, they are better off funding private efforts to sequester the carbon emissions at a fraction of the costs of the proposed programs. Positive sequestration solutions in The Program will win over the skeptical as they solve other problems on which people agree, like soil loss. There are many, many so-called experts asking for funding, but real impact for the long term and collateral impacts are not their concern. As farmer/rancher and engineers, our focus and motivations are for holistic solutions that have positive collateral impact, and for the long term! We need to fund projects that make the soil more fertile and change agricultural practices to soak up more carbon through increased plant growth and keeping living plants on the soil as much as possible. This can all be done through the initiatives and innovations in The Program without the huge sacrifices touted about consistently by the critics.

How many times have you heard about the bad effects of eating red meat and how bans or shutdowns should be in put in place? Unfortunately, we have heard these demands from the very people often flying private planes and traveling internationally, thus bumping up the CO2 level more than from any effect of what you are eating for dinner tonight. Have you ever wondered how many hamburger dinners would equate to one international flight's carbon emissions?

The Get Real Alliance creates innovations to support America becoming carbon-negative, but also with a heart of compassion for the countless low-income people who can't afford any increase in the cost of their basic necessities. With inaction to improve soils worldwide, high-quality food could become harder and harder to find. It is important to act fast regarding Earth's soil degradation for several reasons. As the quality could become more challenged, the costs of producing natural and nutritious foods are increasing more and more; at the same time, we are already seeing rising costs for our basic gas and energy needs. The higher energy costs will be a direct result of legislation creating more and more limits on production and taxes on the everyday producer and distributors.

Added bureaucracy will increase the costs for the average consumer and make cheaper items unavailable altogether in some cases; however, many things can be done with very little disruption and without government interventions. If you follow the www.getrealalliance.org website and subscribe to the newsletter, you can stay abreast of the new ideas as we continue to grow and advance in impact.

Solutions Within the Solutions

There is a lot of talk about how bad things could be at the end of the century as far as climate is concerned, but there is very little discussion on the big issues with real and more imminent danger for this planet. There is not much talk at all about how we are seeing much of the world's topsoil being depleted and how it could be gone by the time we have to really worry about the so-called "global warming" topics of the day. People are not talking much about how our deserts are growing bigger every moment either. By making enrichment of our soil a priority, the CO2 levels will fall, and so much additional benefit will be realized in the health of our planet and our own personal health for decades to come.

There are so many incredible opportunities in America, as we have a large amount of good soil left and huge potential to sequester more carbon than we emit with The Get Real Program's proposals. Making the world carbon-negative will eventually take a global effort, but it can be done affordably. Progress is imminently available to America today! America needs you to help drive the message to greater prosperity and sovereignty in our resources.

The Costs

Big progress in our way of life is not free, despite the occasional irrational thinking that there is such a thing as a "free lunch." The "lunch" always comes at a real cost to someone somewhere. Even the miracle of photosynthesis needs soil nutrients to do its job of creating biomass and food from air and water, with sunshine as the power source. It takes money to change the world for the better, and the best way to solve the climate crisis is to invest in private global efforts and the innovative solutions for real change as are outlined by The Program.

We can lower America's carbon footprint dramatically with a phase-out of coal in favor of more natural gas backing up solar photovoltaic most of the time. This can be paid for through a reasonable usage fee added to producers of carbon-based energy sources. The Get Real Program advocates for a rapid phaseout of coal, which is the most carbon-emitting of all energy sources; however, certain politicians' energy plans generally keep coal in use for political reasons. Balance in our approach is key!

So, exactly what is the plan? Keep reading…

THE PLAN FOR AMERICA SERVES AS A MODEL FOR THE WORLD

The Program has a plan that will make America carbon-negative. Yes, that is the goal… well, that is what will happen when the real problems in America are solved. The real problems that need to be solved today are degradations and underutilization of our soil, oceans, forests, and grasslands. The advances with our natural resources will create carbon sequestration opportunities that will compound until we become carbon-negative.

You may not feel the impacts from the changes we can make in a day or even a year, but farmers and engineers know it will just be a matter of time. It is confusing when depletion of key resources connected to our food supply and other issues highlighted in this book are not what we hear on the news each and every day as talk about climate change and global warming is repeated over and over. As we know, that is where the money warps the narrative and drives the outcomes. This is where The Get Real Program can step in and refocus with real solutions… if you help!

The Get Real Program Difference

Current mainstream climate thinking is about trying to minimize supposed temperature rise due to CO2, assuming that not much can be done to actually reduce the CO_2 level. Scientists are amazed that less than half of man's emissions stay in the atmosphere despite all the carbon-emitting agricultural and forestry practices. With very doable changes in agriculture and forestry, along with restoration of marine wetlands, we can solve the issue of rising CO_2 levels; in fact, we will need to be careful not to lower them too much!

It is proven in greenhouses that growth is boosted by higher levels of CO_2, and many add CO_2 to greenhouses. People have become so fixated on thinking of CO_2 as a bad thing that they forget that it is a limiting nutrient for plants if all other nutrients are available and that lowering it will reduce food production. The Program advocated by The Get Real Alliance is a lot more concerned about the soil and its condition as fixing the soil will also fix CO_2 levels. See, it is a win-win!

This final chapter has raised many issues related and unrelated to climate change and the ecosystem, but we really need to think holistically about the total system and not just single issues. Putting the climate issue behind us will let us focus on the other urgent needs and challenges. We need to take action to save the vital soil and Earth nutrients, starting with the process of remineralization. Turning the CO_2 curve from upward to downward is possible with a global sequestration effort, but America can make a difference in slowing the rise by itself and set a good example for others.

It is great news that we can solve the issue of rising CO_2 without stopping the use of oil and gas. This approach opposes the mantra of the climate community that gives no credit to remineralization and regenerative agricultural practices. It will take a serious effort to really remineralize the Earth – one that is far beyond the initial program proposed in this book, but the benefits would be enormous. The catastrophe that will come with a growing loss of topsoil and depletion of plant nutrients is very slow-acting but stands to kill far more people over time than something like COVID-19.

If the current trends continue, the world could be unlivable at some point in the coming centuries. It is unknown what the effect of rising CO_2 will be over time but stopping it in a productive way through soil sequestration is a win-win! Let's create a world of healthy living soil that takes CO_2 levels down and creates nutritious, abundant food.

YOU MAY WONDER HOW WE CAN HAVE THE GREATEST IMPACT?

A. Invest in our soil and other regenerative and remineralization efforts, including restorative and regenerative agriculture.
B. Encourage more realistic and better forms of renewable energy that have less collateral damage, unlike solar photovoltaic with all its issues:

    a. Rapidly build new natural gas generation for backup to innovative energy solutions and to replace coal as fast as possible.

3. Be good stewards of natural resources while also maintaining our sovereignty and securing our food supply:

    a. Encourage efficiency in everyday living.
    b. Maintain quality of life and avoid major changes in the way we live, but with tremendous impact.

The alternative to the focus above will slowly become apparent as a risk to our safety, health, and sovereignty. Yes, you may not easily see and feel all of the degradations of our natural resources impacting you today, but that is only because the process happens slowly over time. You may attribute the changes you feel over time to old age, but that is only partially true. It is not like the nutrients were sucked out of the soil at one time, suddenly stealing the nutrients in the food supply overnight; it is happening slowly. Therefore, the effects are slight and difficult to notice. We don't know that those things could kill us if soil health is not managed by remineralization and overall better management of our natural resources. It should be engineered so that they can deliver the utmost nutrition and a healthier better future.

Let's dive a little deeper...

INVESTMENT IN OUR SOIL AND OTHER REGENERATIVE AND REMINERALIZATION EFFORTS, INCLUDING RESTORATIVE AND REGENERATIVE AGRICULTURE

As mentioned, we take for granted one of the Earth's most prized resources... soil. It is imperative that we fight the trends of today and show the value of soil, through remineralization. If we do so, we will create the most effective and impactful action yet to fight the so-called global warming and climate change narratives of today. Improved soil is one major solution to a more comprehensive approach. The ability of soil to sequester carbon is the key... we just need to feed the soil to create the greatest impact on the environment.

LET'S FOCUS ON OUR SOIL

Remineralization for Carbon Sequestration Has Compounding Effects

Man has a role in the degradation of our soil, but we have the power to make changes that can turn it all around. We must treasure our soil, constantly nurturing it and cutting out the bad practices of recent years... these include the use of chemical fertilizers and other damaging processes like leaving soil unplanted. The Founder has personally witnessed, as farmer/rancher of the land, the incredible impacts of the solutions we are bringing forth to restore our soil through efforts like rock dust remineralization. Without rock minerals, life cannot exist. Rock dust remineralization increases photosynthesis and nourishes soil life, which leads to more carbon sequestration. We have to cut out the toxic chemicals killing our nutrients in the food supply by investing in rock dust for remineralizing the Earth. Using nature to advance nature is always a good solution.

Soil that is leached and has year-round primary crop growth gets very demineralized, and many countries to the south of the U.S. suffer low agricultural productivity in areas where there is no volcanic soil. Spending money on remineralization with basalt rock dust and other amendments can sequester carbon and boost income with more food production. This economic boost will help diminish the need to emigrate to more developed countries.

Rock dust is not the only solution for filling our soil with rich nutrients so that we can eat a nutrient-rich diet that will give us increased health. Biochar can be loaded with nutrients, preventing them from leaching away as they often do now. It can also play a role in both sequestering more carbon and creating an environment within our soil where life can flourish. Throughout history, biochar has been utilized successfully to revitalize land and areas of water. The potential is unlimited. With our support, farmers can incorporate the ever-available biochar into their farming practices, and we will all benefit!

As part of The Program's innovative thinking on solutions, there are ways to change the way biochar is made for the better. This will make it even easier for biochar to be part of a comprehensive solution. All soils can use biochar to improve water and nutrient retention and improve water infiltration, but many areas don't have the available wood to make it. The methods envisioned and detailed further on www.fullofideas.com have real promise to advance the availability of biochar for more broad use.

Looking forward, as biochar becomes available to the broader areas, The Program has solutions for distribution that are quite innovative and forward thinking. Some ideas shared in the chapter on biochar are provided to get us thinking about future innovations in a new way and to stir conversation. It is a must-read chapter! Let's bring everyone's thinking to the next level together! Join us!

Look No Further at Cattle Reductions

Cattle can be part of the solution for our climate woes, rather than being "the problem." Allan Savory has published significant amounts of interesting material describing innovative and proven approaches to grazing cattle as part of the solution. Everything exists on this Earth for a reason and to support life at its best, and cows are not any different. Make sure you visit the chapter on soil to learn more about how cattle can be part of the solution.

Again, we have to look at man's farming practices, including the many ways that corners have been cut to keep up with the neighboring farmers and to support families through tough competition... driven somewhat by the unfair business practices in pricing cattle. We need to pay farmers and ranchers more to be sustainable holistic managers instead of just allowing them to struggle to keep their business alive. We can use our lands and our cattle to drive advances in the soil's nutrient content to create a stronger food supply system full of more nutrients, while also increasing the soil's ability to sequester carbon at dramatically increased rates.

More Life in Our Soil: Cover Crops

Cover crops preserve the soil for the future. Cover crops can grow continuously between seasons, which can benefit and profit the farmer. For instance, if you are growing corn and then also grow legumes as a cover crop that makes nitrogen; that cuts your fertilizer bill.

Cover crops help in two ways. First, the growing crop sequesters carbon in its biomass. But most of that carbon returns to the atmosphere, only to be reabsorbed by the following crop. With more cover crops planted in both the northern and southern hemispheres, there would be greater carbon sequestration continuously. The second advantage provided by cover crops is that having living roots feeding soil microbes, insects, and animals prevents them from consuming other soil carbon compounds and primary crops for food. They continue building organic compounds that can improve soil quality and permanently keep carbon in the soil. Good use of cover crops provides a net gain for farmers but they have startup costs and require change from traditional multi-generation farming methods that will take multiple seasons to fully implement.

An estimated 3.5 tons of extra carbon and prevented soil oxidation per year per acre for 300 million acres, equates to 1 billion tons of carbon sequestered or not emitted to the atmosphere. Investment and incentives to support healthier farming practices will have longevity and big payoffs when cover crop planning is implemented broadly.

There Is So Much More to Our Soil

This book could focus on soil alone, as there is so much to share and many possibilities to explore. It is at the core of our comprehensive solution on the climate crisis in the news today, due to its incredible ability to help solve for rising carbon levels. There are many other levers to pull in regard to solving for the rising CO2 level, but again, the REAL problems will also be solved by addressing the climate crisis with these very real solutions... dying forests, declining viability and sustainability in our soil resources, challenged ocean and lake waters, and more.

LET'S MAXIMIZE OUR GRASS

We hear a lot about tree planting, but how often do you hear about grass as one part of the solution on climate change? When thinking about how grass is part of the solution, we have to first look at our cow and grass connection. Yes, cows may again be part of the solution, rather than the problem – despite what we constantly hear in the news.

Cows are walking hosts for a wide variety of organisms that can create complex proteins from simple sources of nitrogen and carbon. All of those body functions are actually life-nurturing, which is eventually also lifesaving. The very methane that activists yell about, in cattle burps, can advance our nutrition, as bacteria that produce methane are all part of a larger cycle of methane-eating bacteria that are present around the world and help support and sustain our life. This ultimately leads to more nutrition in our soil that feeds our grasses... the grasses that feed our cows that ultimately feed us. See, there you go, I feel better already just thinking about eating that steak!

We cannot talk about grasslands without mentioning the dwindling dung beetle. When we say we need to manage and preserve this asset in America, we mean from top (what we can see) to the bottom (what is underground). Man's practices are driving our grasslands into infertility. We see the devastating impacts of our chemical fertilizers on nature by the declining presence of the dung beetle, which plays a critical role by spreading out and burying cow manure to fertilize the ground.

No solution would be complete without looking at the devastating impacts of the wild horse herds that are also playing a role in destruction of the grasslands. If left unchecked, the great western habitats will be destroyed to a level that will not allow for an animal to live and thrive there. Again, we look to the federal government's lack of management of the land and wonder why they are not looking to their own resources to be part of the solution for remineralizing our grasslands and other management practices for increased carbon sequestration throughout America... all within their direct control.

Holistic Grazing and Improvement of Grassland

Grasslands are abundant but decline if they are not grazed or are improperly grazed with traditional methods. The federal government is a landowner of massive lands across America, but BIG government has not resulted in BIG advantages in the management of the grasslands. The key to healthy grassland is to stimulate herd ruminant activity under pack predator pressure, rather than ignoring it. Our grasslands need to be preserved and maximized, as part of the climate crisis solution.

Holistic grazing is the most effective way to upgrade grassland and fight their loss to desert. It is perhaps the cheapest way to sequester a lot of carbon but takes a big mind change in the way we manage grazing animals. We need to initiate a movement to motivate herdsmen to follow holistic grazing methods, which requires more people to keep the animals moving to new pasture and allowing the grazed area to recover fully before being grazed again.

Around the world, herdsmen aren't well paid, so they engage in labor-saving practices that destroy the grassland. They may not even know what they are doing to the Earth and our planet by their practices. We need to get the advanced, sustainable approaches taught globally in order to effectuate change, especially in economically challenged countries.

The great news is that far more animals can be grazed utilizing holistic methods, so the food supply is increased overall – not to mention the nutritional advances! By getting this book out to the masses, we can share the message that solutions like this exist and drive awareness of the experts who have proven methods, like Allan Savory. Combining remineralization with holistic grazing makes the grasslands a carbon sponge. Wow, that sounds good! Doesn't it?

Prairie Grass

We don't know about you, but it makes us sick to see waste, especially when it is millions of acres of land. Maybe it is our age, but it really bothers us that the federal government has so much land just sitting there wasted when it can be part of the solution to our climate concerns and the health of our nation. Why are we funding

China to create our energy solutions and fight climate change, when we already have a dormant asset that is ready and waiting to be utilized? Why are we not investing more in our own sovereignty? Why are we looking to our competitive countries, thinking that they have our best interests at heart when they are pushing solar photovoltaic and wind turbines? Do you think they want to set us up for success with our climate and make us stronger, or weaker... and more dependent? We digress...

We simply cannot do the area of prairie grass justice by summarizing it, so we invite you to read all of the interesting details laid out further in the chapter on grass. This is an area where old is now new, as we need to go back to proven techniques that can produce a great deal of grass-fed beef, energy, and better soil life, and can sequester carbon with the prairie grass-enhancing inventions like those at www.prairiegrasssystems.com.

The Get Real Program has exciting and innovative solutions that have not been revealed to the masses. By investing in high-intensity prairie grass management, we can prevent soil erosion and increase carbon sequestration. It is so exciting to know that we have an opportunity for great impact now and we, as taxpayers, actually own it! All we need to do is manage what we have... well, we and the federal government, that is. We don't need to look to China and Russia. The solution is in our backyard, here in the U.S., literally.

By investing in innovative management techniques of our prairie grass, we can also increase food and energy production while we sequester carbon. The Program includes creative thinking and advances in how to harvest tall grass and grow polyculture systems. Please make sure you visit www.fullofideas.com for more information on the prairie grass harvesting system and other inventions available to us to create real impact.

LET'S CLEAN UP OUR FORESTS

Forest, Trees, and the Desert Shrubs

Carbon sequestration is the answer to our rising CO2 levels, and healthy forests are essential for carbon sequestration. This is why you see so many volunteer organizations supporting and rewarding tree planting. Planting is definitely lifesaving, but there is so much more to our trees than just planting one.

It is imperative that we all take a close look at the solutions of The Get Real Program that are laid out in this book in great detail. We must not miss one opportunity to use our natural resources to create real impact. As part of the solution, The Get Real Program proposes a comprehensive approach to our forests, trees, and desert shrubs. This includes:

- Innovative approaches to the way timber is harvested, especially on federal lands.
- Adoption of The Program's aerial harvesting and forest management techniques to minimize the destruction of trees that aren't being harvested, with reduced fire risk.
- Remineralizing new and old trees for increased growth and health.
- Challenging man's current practices that are harmful to the forests, including reducing clear-cut logging to avoid the massive loss of topsoil that it causes, and much more.
- Generally maintaining growth of our forests by not only planting trees, but also investing in the cleanup of our deadwood and other fire ignitors, creating biochar and better fertility and biodiversity.

Fire Igniters

We hear so many devastating stories in the news about wildfires, but most of them are avoidable with proper tree and forest management practices. Private property owners know the devastation that can be caused by the lack of proper tree management, but the federal government does not consider it a priority to manage our forests in a way that would reduce the devastation that we see over and over, year after year, when the fires start and take off for many, many miles with devastating consequences. The solutions are out there, but it will take the federal government to be part of the solution for real change to happen.

Our Trees and Forests Are Dying... This Must Stop, or We Will Not Survive

In addition to the massive wildfires, insects and disease are killing our trees. We must make our soil and forests more of a priority or we will have devastating impacts. This is another area where people are not looking at the right solutions or are advancing interests that only slow the problems, but don't solve them. We are on an unsustainable path with the destruction of our forests and trees, which are required for our very breath. What would you do to have years added to your life? What will you do to insure it?

There are many ideas thrown out in this chapter, so we believe it is another must-read for innovative thinkers. FOI Group, LLC has applied modern age thinking to the problems deep within our forests, like airborne lighter-than-air ships for operation centers and other similarly innovative ideas that solve for access deep within the forests to preserve the trees along the way to them. It will be exciting to see what advances can be made as we increase the visibility of the many solutions from The Program.

LET'S STOP IGNORING OUR WATERS... OCEANS, REEFS, WETLANDS, AND MORE

When you think of a resource that could sequester way beyond anything we could possibly produce, making us carbon-negative, look no further than the oceans. Utilizing the vastness of our sea waters could create an amazing impact on our fight for thwarting impacts of rising CO2 levels. Let's face it, we are in a crisis with many locations in our sea waters... acidic areas, overfished habitats and otherwise misused and abused parts near shore and overall. America and the rest of the world are standing by as many acres of marine wetlands are being destroyed or degraded. So much can be done with improvements to marine wetlands and reefs to increase carbon sequestration and marine life, not even counting the opportunities to improve open ocean and deeper fisheries.

Ocean and Marine Wetland Improvements

As with our soil, part of the comprehensive solution lies within remineralization. Remineralizing the ocean and marine wetlands with rock dust, as well as bringing other restoration, offers great benefits on a lot of levels. The proven biorock technology really creates and stimulates reefs and other carbon sequestering life. Large-scale remineralization with the right minerals can really increase carbon sequestration and marine growth. We need to make the ocean more productive to sustainably provide more food for a growing population.

Let's Utilize What We Have

When reading the chapter on our oceans and water, you will learn of some interesting techniques and inventions that can create greater opportunities to utilize our water as part of the solution. There are many fun solutions like creating floating lagoon islands, using wave energy for power, building wind-powered ships, re-imagining our harbors beyond tradition, and so much more. We envision a new nautical tomorrow and would love for you to join us for the important voyage! Hop in!

CARBON SEQUESTRATION AS A SOLUTION

It is important to think of carbon as a part of life. We don't want to destroy carbon. When we talk about the comprehensive solutions to increased carbon dioxide, the answer is to sequester it for its greatest use, not remove CO_2 altogether. CO_2 is two-thirds oxygen and one-third carbon, so that translates to removing three times as much as carbon alone by weight. As carbon is sequestered, more oxygen is available for the atmosphere, which is a benefit to us humans! Some measures do not result in permanent removal from the carbon cycle but are cheap processes per ton, and if sustained, do lower atmospheric levels until reversed – *if* they are ever reversed. Others, such as biochar, are more expensive but their benefits are nearly permanent.

Some ideas in this book would create expensive programs that don't sequester much carbon in the first year but have lasting valuable benefits that can grow over time. We advocate for starting where we can and growing as we are able to demonstrate real impact. Things like holistic grazing and cover crops take money and a management mindset change to get started, but they produce substantial economic gain once they're implemented that will further improve rural life and profitability. This estimate is pretty conservative but shows the incredible potential that nature has to reverse atmospheric carbon rise.

There have been numerous reports generated that offer much lower carbon sequestration numbers, but there is a widespread lack of knowledge of benefits of remineralization or even holistic grazing. Despite decades of success restoring grasslands, Allan Savory's methods are still not widely adopted.

This proposed Program is just a start on what a national program will do and needs a lot of input and development to optimize it for mass and eventual worldwide adoption.

CLIMATE AND ENERGY

It would be remiss of me to talk about climate change without addressing the elephant in the room... energy. Yes, not only do we have natural resources that can sequester carbon out of the environment, but we can also create real impact through the energy we invest in and use. The question is not just "What energy?" but also the *who* behind the what. Do we want to invest in ourselves or our competitors? Right now, unfortunately, we are investing our resources towards foreign energy by giving China the majority of our solar photovoltaic investment.

Sources of advanced energy options exist…

- Nuclear
- Solar thermal
- Wind
- Wave
- Biogas
- Geothermal
- Hydrogen
- Algae

You will have to read more in the materials on these topics, as they are too broad to give them any real credit in a summary format. It is definitely worth your time to read all the chapters here if you have not already! We always jump forward to reference materials, so you may be just like we are. This is not an area to skip, as it has real-life consequences to your everyday way of life. You will also find other gems of information that will be pertinent to the issues. Don't wait until it is too late to educate yourself, because here in this book we get real, honest, and practical on energy and climate… which is hard to find anywhere else! You can go to www.fullofideas.com and www.getrealaboutclimate.org for more information and sources. Sign up for our newsletters for more updates.

As mentioned throughout the book, our energy planning is all tied up in our political environment of the day. This causes great pause for anyone, as it is hard to trust the headlines and research with so much money invested in special interests and energy marketing schemes. Knowledge of the misguided forces behind many of the initiatives and research so many are depending upon gives us an even bigger incentive to create real change through private, non-governmental means.

A GLOBAL PROGRAM

A global program is needed with a much larger amount spent on remineralization as so much of the world is very badly demineralized, including the precious Amazon. Sharply higher timber growth can play a big part in making the world carbon-negative and we can eliminate world hunger in a sustainable and carbon-negative way with a move to regenerative agriculture and holistic grazing.

The world is a big place with a wide range of economic conditions, energy use, and potential for carbon sequestration. We truly need energy-consuming, high-density countries with little fertile ground to subsidize carbon sequestration efforts in countries with lots of land but very little money.

A PROSPEROUS FUTURE

We all want a prosperous future for many future generations, but many politicians are making decisions based upon what can help them politically now, without regard to any future impact on the environment for the long-term or the enormity of the costs to current and future generations. It is so sad to see many leaders of our country serving as such a bad example for managing our budget and only creating costs that balance out with our income opportunities.

Not only are politicians not supporting and investing in the right solutions in a way that creates a positive impact, but they are also absorbing costly bills from China to support the ineffective approaches. Funding China creates less independence. Is that what we want to support with our hard-earned tax dollars... making China stronger and more prosperous? We say no.

We cannot be prosperous as a country if we invest foolishly in expensive proposals that just slightly slow the rise of CO2 instead of actually lowering levels less expensively. The Get Real Program sees another way to lead America to prosperity, while we gain more and more independence, security, and sovereignty. We are being tested by our competitors and history will look back at this time and may reveal the foolish ways America played into propping up our competitors, while creating increased vulnerability for ourselves. This is not the legacy we want to leave for our grandchildren and their grandchildren. We want to leave this world a better place, full of richness in the soil, a nutrient-dense food supply, and fresh air, and without a huge deficit taking all of their hard-earned profits.

The Program has many innovative solutions that can give you confidence our rising CO_2 levels can be affordably managed in a positive way and so much more. The comprehensive changes proposed are positive, but massive, and will have some strong opposition for those lining their pockets on the wrong solutions or problems. Almost everyone will benefit greatly from the plan, so with your help, a concerted effort will be enacted!

We know that many of the climate lobbying firms are the biggest and strongest in Washington, D.C. They will be a force to reckon with, but The Get Real Alliance is prepared and backed up by facts that are undeniable. The climate lobbyists are focused on reducing fossil fuel emissions in an illogical way that slows the reduction of CO_2 emissions by banning new natural gas plants that could replace polluting coal plants. The coal industry and its unions are fighting for their lives and are very opposed to new natural gas plants that would put them out of business.

The news these days is so full of misinformation, so we hope and pray that you are reading this material to educate yourself in order to discover your own facts on what is happening from an accurate, honest source. The Get Real Alliance Founder's life work has been to do the same and spread the word.

It is not easy to fight the big voices that carry big dollars in our nation's capital. The climate lobby only know the surface of what is happening, and they are spinning it to their advantage. They are not generally farmers and engineers who have put their hands in the soil, who have spent time watching the changes in farming get more and more complex – disincentivizing investment in natural solutions. They are stuck in concrete in the urban Washington, D.C. environment, not in nature. They do not know what they do not know, but we do see a broader context!

Get Real on Politics and the Coal Connection

Politicians are often transparently not for advancing our way of life, but their own. We can see this in recent legislation banning natural gas. It does not take a genius to see the warped conflicts of interest around Washington, D.C. regarding natural gas. Since natural gas drilling and production is not a unionized industry and generally the industry does not support the Democratic party, it is not surprising that it got banned in the new Virginia climate bill put forward by the Democrats there.

The bill is an example of short-term benefits for the core group – the coal industry, with its increased CO_2 emissions continuing for some time with a distant and unrealistic goal of 100 percent renewable power with wind and solar photovoltaic. Virginia has substantial agriculture, so large-scale adoption of digesters producing biogas, coupled with algae production, would be more sensible. But the anti-natural gas provisions prevent this from being a major part of the solution. Passing a law that defers into the distant future key elements that help reduce carbon emissions, while banning immediate reductions to help an industry (i.e., coal) survive even though it is uneconomic, is typical dirty politics.

We need to GET REAL and think holistically about the climate, the Earth, and all its inhabitants. The Get Real Alliance is fervently anti-coal because there are so many negatives – including that it effectively blocks intermittent renewables from being used when available. The environmental cost is so high that paying miners severance and retraining them when natural gas replaces coal is still cheaper and cleans up the environment in so many ways.

The coal industry is an obsolete, money-losing business that is being kept alive purely by politics. Building factories to make the required components for biochar facilities in coal mining areas can provide good jobs for the displaced miners. Being realistic about politics means good compromises have to happen, and it is critical to save the soil and us for the long-term.

Climate Should Be Non-Partisan... Not Corrupted to Advance Special Interests

This book and The Get Real Alliance are not partisan. The solution can be found only in going beyond partisanship and holding both parties accountable for their abuse of nature and of people and resources in agriculture. Your support is crucial to the visions shared within this book becoming a reality, so please join us and promote the cause by sharing this book, the messages within, and the issues with solutions on social media.

To succeed against the strong lobbies of coal unions, agricultural chemical companies, the anti-fuel-tax crowd, the current structure of agriculture, and even some in the climate change and environmental community, The Get Real Alliance will need to present a very strong lobbying effort with a broad base of support. It all starts with intelligent, educated people sharing the truth with facts to back us up... based upon real experience. The status quo is very difficult to change, even if change is urgent and essential for long-term survival.

Uncertain Times

The writing of this book was well underway when the COVID-19 pandemic really disrupted the world, but two important things have since happened. First, fossil fuel got cheaper and then more expensive as supply constraints ensued. Some of this disruption appears to be due to the irrational and destructive focus on the Green New Deal and other imprudent insufficient programs. Second, the pandemic revealed our need to think holistically and be prepared for uncertainty. Frankly, the global focus over a relatively tiny percentage of the world population dying from COVID-19 could be less important compared to the dire prospect of widespread starvation caused by the impending soil loss and desertification. The difference between an unlikely death in the very near future and nearly certain death for much of the world's population in the distant future plays to man's weakness: we can rationalize taking no action, even in the direst of situations, if we aren't the ones who have to pay the price for inaction.

This book is about a positive solution to the climate crisis, but its suggested program addresses other issues. The Program can change the trajectory of CO_2 levels from rising to falling; however, in the long run, more changes have to be made, and they will cost money. Just as in the long run we must come to rely on sustainable energy instead of depleting fossil fuels, so too must we restructure the way we live so as not to run out of other Earth resources in the future.

YOUR CALL TO ACTION STARTS HERE AND NOW!!!

Your help is needed to change the mindset of climate policy advocates away from impractical and harsh proposals to cut carbon emissions to zero and toward positive and productive, sensible increased use of smart energy options that actually make reliable power such as biogas, algae fuel, new wind power-gas turbine designs, wave energy, and geothermal wells. Turning the world carbon-negative while we still use oil and natural gas takes away the urgency to make rash, mandated changes towards unreliable power like some wind and solar photovoltaic designs. We need your support, so sign up at www.getrealalliance.org and on social media and spread the word!

Too often, people get overwhelmed with the big issues in life and think that they cannot personally make a difference, so they do not do anything. They feel stuck and look to leaders to manage the big issues. Even a volume of small issues can be overwhelming or our day-to-day can take our eyes off of the big issues. This is where The Get Real Alliance can help. We can step in and create real impact and change by using your donations to focus action on the most concerning problems of our day. The Get Real Alliance has a laser focus on our planet's sustainability and our food supply of the future.

There are good organizations working to build awareness that need support. See www.thecarbonunderground.org for ways to help. Another small non-profit that has some good information is www.soilcarbonalliance.org. And www.remineralize.org is doing great work and really needs your support. This book is held by www.getrealalliance.org, which is a non-profit 501(c)(4) that can engage in political education. **Donations to Get Real Alliance are not tax deductible; however, also, the donations and donors are not disclosed to the public.**

We need you! It will take strong, educated advocates like you to make a real difference in changing the climate proposals from ineffective, expensive proposals to proposals that can solve the issues. Donations are not tax-deductible, but they can really make a difference politically, which is where the change needs to happen. We travel to promote soil carbon sequestration and would be happy to speak with your group of 40 or more people if possible.

Now is a time for action to start really sequestering carbon by getting the carbon sequestration program enacted. Please join and support www.getrealalliance.org to take the next step. If, after reading this book, you are unconvinced of the urgent need for The Program, please review the source material and go to the Alliance website for more links and articles.

Because there are some very dangerous irrational proposals surrounding climate issues, this book was expedited. We've included information about some inventions that are at a very early stage of development and analysis. Hopefully, you can see which areas are speculative and which are proven successes such as holistic grazing, cover crops, biorock, biochar, and remineralization.

Lastly, book sales and donations to The Get Real Alliance will fund study, promotion, and more research and development on the inventions and topics.

Let's Get Real and work hard to make the world a better place. For posterity!

"The only thing that is necessary for evil to triumph is for good men to do nothing."

— Edmund Burke

REFERENCE LIST

Savory, Allan, with Butterfield, Jody, ***Holistic Management: Third Edition: A Commonsense Revolution to Restore Our Environment,*** Washington, DC: Island Press, 2016. Cited - Chapter 2; page 32 and numerous other places.

Remineralize the Earth (RTE) is a 501(c)(3) non-profit organization based in Northampton, Massachusetts, and founded in 1995 by Joanna Campe. https://www.remineralize.org Cited - Chapter 3; page 39.

Hamaker, John D., Weaver, Donald A., ***The Survival of Civilization Depends Upon Our Solving Three Problems: Carbon Dioxide, Investment Money and Population - Selected Papers of John D. Hamaker,*** Hamaker-Weaver Publishers, First edition, June 1, 1982. Cited - Chapter 3; page 44.

Edited by Goreau, Thomas J., Larson, Ronal W., Campe, Joanna, ***Geotherapy: Innovative Methods of Soil Fertility Restoration, Carbon Sequestration, and Reversing CO_2 Increase,*** Boca Raton, FL, CRC Press, Taylor & French Group, 2015. Cited - Chapter 3; page 57.

Scheub, Ute, Pieplow, Haiko, Schmidt, Hans-Peter and Draper, Kathleen, ***Terra Preta, How the World's Most Fertile Soil Can Help Reverse Climate Change and Reduce World Hunger,*** Vancouver BC, Canada, Greystone Books, 2016. Cited - Chapter 4; page 62.

https://www.allpowerlabs.com/carbon Cited - Chapter 4; page 64.

Kinsey, Neal and Walters, Charles, ***Neal Kinsey's Hands-On Agronomy: Understanding Soil Fertility & Fertilizer Use,*** Acres U.S.A., Greeley, CO, 2013. Cited - Chapter 5; page 86.

https://www.acresusa.com Cited - Chapter 5; page 87.

https://www.megafood.com Cited – Chapter 5; page 88.

Hensel, Julius, ***Bread from Stones,*** Greeley, CO, Acres U.S.A., 1991. Cited in Chapter 5; page 90.

Mays, Daniel, ***The No-Till Organic Vegetable Farm: How to Start and Run a Profitable Market Garden That Builds Health in Soil, Crops, and Communities***, North Adams, MA, Storey Publishing, 2020. Cited - Chapter 5; page 102.

Schwartz, Judith D., ***Cows Save the Planet: And Other Improbable Ways of Restoring Soil to Heal the Earth,*** Hartford, VT, Chelsea Green Publishing, 2013. Cited - Chapter 6; page 114.

Wikipedia contributors. ***Methanotroph***. Wikipedia, The Free Encyclopedia. October 21, 2021, 00:23 UTC. Available at: https://en.wikipedia.org/w/index.php?title=Methanotroph&oldid=1050982689 Cited – Chapter 6; page 120.

https://www.nationalgeographic.org/encyclopedia/rain-forest/ Cited - Chapter 7; page 142.

https://timberupdate.com/blog/thinning-timber-can-earn-350-per-acre-and-increase-annual-harvest-by-25/ Cited – Chapter 7; page 147.

https://www.soilcarbonalliance.org/2017/02/01/regenerative-development-to-reverse-climate-change-quantity-and-quality-of-soil-carbon-sequestration-control-rates-of-co2-and-climate-stabilization-at-safe-levels/ Cited - Chapter 8; page 163.

Bristow, L. A., Mohr, W., Ahmerkamp, S. and Kuypers, M. M. M. (2017). ***Nutrients that limit growth in the ocean***, ***CURRENT BIOLOGY,*** 27(11), R474-R478, doi:10.1016/j.cub.2017.03.030. Cited – Chapter 8; page 164.

Moore, Mark C, ***Diagnosing Oceanic Nutrient Deficiency,*** The Royal Society 374.2081, 28 November 2016. https://royalsocietypublishing.org/doi/full/10.1098/rsta.2015.0290 Cited Chapter 8; page 165.

Edited by Goreau, Thomas J., Trench, Robert Kent, ***Innovative Methods of Marine Ecosystem Restoration***, CRC Press, Taylor & Francis Group, Boca Raton, FL, 2013. Cited - Chapter 8; page 90 and Page 174.

Goreau, Thomas J.F., ***Biorock Technology: A Novel Tool for Large-Scale Whole-Ecosystems Sustainable Mariculture Using Direct Biophysical Stimulation of Marine Organism's Biochemical Energy Metabolism,*** http://www.globalcoral.org/2018-international-summit-on-fisheries-aquaculture/ Cited - Chapter 8; page 175.

https://globalcoral.org/ oldgcra/reef restoration using seawater.htm Cited - Chapter 8; page 175.

Shellenburger, Michael, ***Apocalypse Never: Why Environmental Alarmism Hurts Us All***, Harper, illustrated ed., June 30, 2020. Cited - Chapter 9; page 185.

Martin, Chris, ***Wind Turbine Blades Can't Be Recycled, So They're Piling Up in Landfills***, Bloomberg Green, February 5, 2020. Cited in Chapter 9; page 187.

Forstchen, William, ***One Second After (A John Matherson Novel, 1)***, Forge Books, New York, NY; November 2009. Cited - Chapter 9; page 207.

Svensmark, Henrik and Calder, Nigel, ***The Chilling Stars: A New Theory of Climate,*** Icon Books Ltd., London, 2007. Cited - Chapter 10; page 222.

https://www.kinseyag.com Cited - Chapter 11; page 253.

Multiple references were from the Allan Savory YouTube TED video: ***How to green the world's deserts and reverse climate change | Allan Savory***. https://www.youtube.com/watch?v=vpTHi7O66pI

A great source for many good books is https://bookstore.acresusa.com which has many books by Dr. Albrecht and others. There are a number of books on biochar and soil minerals and testing. The founder has learned much from their very good monthly magazine. They have been leaders in pointing to sources and materials on Regenerative Agriculture, including:

Albrecht, William A., Ph.D., ***Albrecht's Foundation Concepts, Vol I,*** Acres U.S.A., Greeley, CO, 1975, 2011.

Albrecht, William A., Ph.D., ***Albrecht on Pastures, Vol. VI,*** Acres U.S.A., Greeley, CO, 2011.

Albrecht, William A., Ph.D., ***Albrecht's Soil Fertility & Human & Animal Health, The Albrecht Papers, Vol. VIII,*** Acres U.S.A., Greeley, CO, 2013.

(Dr. Albrecht is cited beginning in Chapter 5).

Other suggestions for future reference:

Savory, Allan, with Butterfield, Jody, ***Holistic Management: A New Framework for Decision Making,*** Washington, DC: Island Press, 1998

Butterfield, Jody, Bingham, Sam and Savory, Allan, ***Holistic Management Handbook: Third Edition: Regenerating Your Land and Growing Your Profits,*** Washington, DC: Island Press, 2019

Svensmark, Henrik and Calder, Nigel, ***The Chilling Stars: A New Theory of Climate,*** Icon Books Ltd., London: June, 2008.

CPSIA information can be obtained
at www.ICGtesting.com
Printed in the USA
JSHW030847140422
24908JS00002B/4

9 781956 914245